CONTENTS

Acknowledgements

The Working Party for this third edition gratefully acknowledges the contributions made by the current and previous Maintenance Practice Committees.

Composition of the Working Party for the third edition:

Chairman   Victor E Michel FCIOB

Members   Stuart M Bradburn FCIOB
Derek J Cripps FCIOB
Leslie A Etherington FCIOB
Herbert Graham MCIOB
Michael H Hall FCIOB
Edward Hankin MSc MCIOB
Clifford J Pratt MCIOB
Christopher E Smith FCIOB
Harry S Staveley FCIOB
Robert F Stevens MEng PhD
Norman Thomas *FCIOB*
William Wilkie BSc *MCIOB*

Composition of the Maintenance Practice Committee responsible for the second edition:

Chairman   R.J. Bushell, MBE, BSc, FRICS, FCIOB

Members   N.R.G. Clarke, MSc, MCIOB, ARICS
D.A. Cutting, MCIOB, MBIM, FFB
H. Graham, FIMunBM, MCIOB
L. Hill, FCIOB
N.R. Marsh, MCIOB, MBIM
E. Shneerson, FCIOB
H.S. Staveley, FRICS, FCIOB
J.E. Trill, MCIOB
W. Wilkie, MCIOB

Secretary   F.C. Lettin

Composition of the Maintenance Practice Committee responsible for the first edition:

Chairman   E. Bampton, FIAS, FCIOB

Members   N.E. Ellis, MCIOB
R.J. Bushell, BSc, FRICS, FCIOB
H. Hollingworth, ARIBA, FCIOB
T.R. Nicholson, MCIOB
T.R. Norton, MCIOB
R.N. Prentice, MCIOB
C. Sumpter, MCIOB
J.E. Trill, ACSI, AMBIM, MCIOB
R.A. Wisby, MCIOB

Secretary   P.A. Harlow, AIM, MIInfSci

Terminology

The following are the meanings of the principal specialist terms used in the guide. Where appropriate they conform to BS:3811:1976 *Glossary of maintenance terms in terotechnology.*

ACCEPTABLE CONDITION. As-built drawings prepared on completion of the building, or sections of the building, which show the details as actually found on site. These may differ from the original design drawings.

ATTRIBUTE. The overall characteristics or features of a building, such as its size, general internal and external layout and principal constructional details. The term is increasingly used in connection with condition assessments, where if employed loosely, it may be confused with elements and components (qv).

BONUS DRIFT. Bonus drift occurs when bonus earnings are allowed to increase above the level justified by, and appropriate to, the achieved output. It arises when management fails to maintain tight control over the day-to-day operation and calculation of the bonus scheme and to update time values to changing circumstances.

BUILDING AUDIT. The regular and structured examination of industrial buildings and associated facilities (in relation to the industrial activities conducted within) that:

> identifies elements of the building or facilities where improvements to the standards or performance which obtain would be appropriate to the needs of the user;
> provides a schedule of such improvements together with cost indications for the improvements concerned;
> quantifies the financial return to the user/owner against the investment in specific improvements;
> identifies benefits (environmental, image, staff morale etc) where financial quantification of returns from investment in improvements is a matter of judgment;
> indicates the appropriate route for the realisation of the improvements – professional advice and services.

BUILDING OWNER. The individual, group of people or corporate body who owns the premises comprising the estate.

CLEAN. To reduce contamination to an acceptable condition.

CLIENT. The building owner or the person(s) in whom responsibility and accountability for the maintenance function may be vested. A distinction has to be drawn between the building owner, as client for whom maintenance work is carried out by contractors, and the tenant or occupant, who may be more conveniently referred to as the 'customer'.

COMPONENT. A specific item within the fabric, services or finishes of a building whose breakdown or decay would lead to a failure or progressive deterioration in performance in the element of which it forms a part.

CONDITION ASSESSMENT OR SURVEY. A technical examination of the condition of properties, both internally and externally, with a view to identifying the principal components where repairs or renewals are likely to be required over a given period of time.

COST-IN-USE. The total cost of providing a running premises under the following four groups of expenditure:

Group 1: The capital cost as an annual sum (amortisation), plus the loss of interest on the capital or the rent, plus ground rents, etc, plus the rates and insurance on the premises.

Group 2: Utilities: gas, water, electricity and heating.

Group 3: Repairs and maintenance including alterations and additions to the fabric and engineering services.

Group 4: Ancillary services.

DAYWORK. A method of reimbursing the cost of repairs, which may be too complex or too small to warrant valuation by measurement. Labour is paid for at a daily or hourly rate and materials and plant at prevailing rates, plus a percentage for overheads and profit.

DEFECT. An unexpected deviation from requirements which would require considered action regarding the degree of acceptability.

ELEMENT. The major parts of a building that are recognisable as performing a key function within the structure, services, finishes or surrounds into which they are built.

EMERGENCY MAINTENANCE. Maintenance which is necessary to be put in hand immediately to avoid serious consequences.

ESTATE OWNER. A person, group of people or corporate body owning land and/or buildings comprising an estate.

EXAMINATION. A comprehensive inspection supplemented by measurement and physical testing in order to determine the condition of an item.

FACILITIES MANAGEMENT. Facilities refers to property where people are accommodated and work or where an organisation conducts its business. Management concerns all aspects of providing, maintaining, developing and improving those facilities.

FAILURE. The termination of the ability of an item to perform its required function.

INSPECTION. A careful and critical scrutiny of an item carried out without dismantling by using one or more of the senses of sight, hearing, smell, taste and touch.

LIFE CYCLE COSTING. A technique for assessing the total cost of an asset over its operating life including initial acquisition costs and running costs.

MAINTENANCE MANAGER. A generic term covering all those responsible for the organisation of maintenance.

MAINTENANCE MANAGEMENT. The organisation of maintenance within an agreed policy.

MAINTENANCE MANUAL. A manual, usually loose leaf, containing all essential information about the construction of the building, together with all the necessary sources of information for its proper maintenance.

TERMINOLOGY

MAJOR REPAIRS. Work similar in scope and manner of execution to planned maintenance or programmed repairs, but arising mainly from the necessity to rectify defects early in the life of a building caused by poor design or workmanship at construction stage.

MINOR WORKS. Isolated items of work, often to unoccupied premises, that can be undertaken more cost-effectively as a single contract than on a jobbing basis and cannot be delayed for inclusion in a major repairs programme.

PACKAGE DEAL. An arrangement between the client requiring a new building and a contractor to design, erect, commission and hand over the completed building on agreed financial terms.

PLANNED MAINTENANCE. Maintenance organised and carried out with forethought, control and the use of records to a pre-determined plan.

PLANNING. The determination in advance of the jobs, methods, materials, tools, machines, labour, timing and time required.

PREVENTIVE MAINTENANCE. Maintenance carried out at pre-determined intervals, or to other prescribed criteria, and intended to reduce the likelihood of an item not meeting an acceptable condition.

QUALITY ASSURANCE. An objective demonstration of the builder's ability to produce building work in a cost effective way to meet the customers requirements.

REPAIR. The restoration of an item to an acceptable condition by the renewal, replacement of mending of decayed or damaged parts.

RESTORATION. Maintenance actions intended to bring back an item to its original appearance or state.

RUNNING COST. The cost necessary to provide the facilities for the main function of the occupier of a building, eg, provision of heating, lighting, power, water, gas or other utility services. It includes the cost of janitors, porters and cleaning staff, window cleaning and laundry, together with other ancillary services.

SCHEDULE OF PRICES (OR RATES). A list or schedule of items of work specified in detail against which a contractor puts his rate and which can then be measured on completion and the total costs agreed. Alternatively, the rates against each item may be given by the maintenance manager. The contractor then or deducts adds a percentage against each trade section to cover his overheads, travelling costs and profit.

SCHEDULE OF RATES. A document or set of documents listing items of repair work or tasks, with provision for the insertion of prices per unit of measurement as a basis for tendering by contractors and payment for work carried out.

SERVICE CONTRACT OR AGREEMENT. A form of contract, accompanied by a schedule of works and specification, used primarily for the regular maintenance of mechanical and electrical equipment such as boilers, lifts and fire alarms, and normally subject to annual renewal.

SERVICING. The replenishment of consumables needed to keep an item in operating condition.

SURVEY. An examination, the written report of which includes a recommendation for action.

TERM CONTRACT. A form of contract used principally for the jobbing repairs in conjunction with a schedule of rates (qv.), where there are variable quantities of work to a number of small but frequently recurring items over an agreed fixed period, usually between one and three year's duration.

Chapter 1

Introduction

Buildings are a nation's most valuable asset, providing people with shelter and facilities for work and leisure. Contributions to this asset have been made by successive generations and its value now runs into many billions of pounds. Apart from their monetary value, buildings form part of the national heritage, the best examples of which must be preserved for future generations. The importance of maintenance can be judged from its relationship to the building industry and to the gross national product (GNP). Maintenance represents 40% of the output of the building industry, and in turn the building industry represents about 10% of the GNP.

No less significant is maintenance management, which requires a variety of skills, technical knowledge, and site experience in order to identify and satisfy maintenance needs. An understanding of modern technology and business management, knowledge of the law as it affects people, property, and contracts, and perhaps most of all, an understanding of people is essential in order to specify remedies, since the vast majority of buildings exist to satisfy the needs of people. A prime mover in the establishment of maintenance management practices must be the building owner and his agents, for they provide the resource and motivation for the caretakers, estates surveyors, building surveyors, engineers and graduates in estates management, who are responsible for the large estates and the maintenance of building stock.

The purpose of this guide is to help the building owner and his staff, by encouraging them to anticipate the working life of their buildings and their present and future use. It will also assist them in specifying a good and cost effective practice in the organisation, management and execution of maintenance works and encourage them to plan maintenance in advance, to avoid the penalties of an ultimate failure.

The guide is further designed to impress upon building owners and designers the need to consider maintenance at the conceptual stage of building design; it should not be a subject for thought after the erection of a building. Consideration must be given at this stage to the building's use and its life cycle, as well ensuring that proper access is designed in to facilitate maintenance in the future. Building users, when acquiring speculatively constructed buildings, must be prepared to make a critical appraisal of future maintenance needs, particularly when purchasing leases on a full repairing basis. By doing so they will be in a position to influence developers to produce the maintenance free buildings of the future. The basic principles which apply to the maintenance of buildings are also presented in the guide. Many relate to all buildings, but because of the broad scope some are more appropriate to larger estates than to the single house, shop, or factory. Proposals are made for rationalising building maintenance for building owners with directly employed labour (DEL), for those utilising contractors to carry out maintenance work, and for contractors themselves who carry out maintenance work. A number of steps can be taken to facilitate more effective management in building and these are set out below and studied in detail in this guide:

MAINTENANCE MANAGEMENT

- Maintaining of sites, buildings, engineering services, service agreements, renewal dates, decorations and costs.
- Working to standard procedures.
- Making greater use of maintenance planning, including service standards, and response times.
- Training managers, supervisors, and maintenance operatives more effectively.
- Examining methods by which work is put out to contract and contract procedures.
- Effecting good lines of communication and dissemination of information.

It must be stressed that the above subjects–and all the information contained in this document–act purely as a guide. They should be used to create tailor-made practice manuals for a particular estate and not as practice manuals in themselves.

No building is immune from the ravages of time and weather and to preserve it maintenance is essential. Often a building represents an organisation's major capital asset, yet it is a paradox that many building owners regard maintenance as an undesirable overhead and accord it a low priority. In times of recession, history has shown that the maintenance budget is invariably the first target for reductions. The deferment of major repair and renewal works accelerates the decay of buildings and brings about a reduction in asset value, resulting in an unnecessary increase in expenditure for future times.

Government has also made a contribution to this neglect, for example, by the introduction of VAT which has directly increased maintenance costs for organisations and individuals who are unable to reclaim the VAT charged. Even where grants are available for the repair and improvement of properties, complexities of law make it difficult for a building owner to establish what might be obtainable.

Maintenance is prevention of failure. It is false economy not to maintain buildings, their engineering services and environment. Building owners are often dilatory in enforcing maintenance clauses in tenancy agreements, thus allowing tenants to escape their obligations until the final surrender of the lease, when often the building is beyond economic refurbishment. Ignoring maintenance can result in serious failure, which involves significantly more resource and greater disruption of occupancy than would be the case if normal maintenance routines were followed.

Maintenance work is often said to cause inconvenience to a building owner or occupier and can frequently disrupt industrial production, but it must be remembered that total chaos can arise from failure; it is necessary to consider maintenance at the appropriate time, allocate funds and plan accordingly.

To a building owner costs-in-use are often of more importance than the initial cost price of the building. Because of the industry's practice of accepting the lowest tender for work, it is quite common for initial savings to be made up by high maintenance costs in the future.

Such practice tends to ignore the life cycle costs of the building in which after-care costs over the years outweigh the savings made on the initial outlay. However, it must be appreciated that to a developer, intending to let a finished building on a full repairing basis, life cycle costs are unimportant.

In consequence it is important for a proper brief to be given to the design and construction teams in the first instance, with particular emphasis on the final building being fit for its intended purpose, thus keeping ongoing costs and defects to a minimum.

If life time cost considerations are ignored, then they will continue to rise with inflation, bringing about rising costs in energy, labour and materials. The need to keep these increases in check is a constant challenge to the maintenance manager.

The effect of increases in the cost of new building work can be offset in part by the introduction of new materials, labour saving equipment and new approaches to design, resulting in higher productivity. However, the maintenance budget, being primarily devoted to existing buildings, does not benefit from these trends. In fact, there is sometimes a positive disadvantage, as some new approaches to design can bring about higher maintenance costs when replacement becomes necessary,

Good design, sound construction and, correct use of materials can reduce maintenance costs without necessarily increasing new building costs, but research is necessary to fill the gaps in present knowledge, as metal cladding, pre-cast panels, GRP panels, and the like, whilst saving initial costs, are known to bring about massive increases in maintenance costs. To be of value, all the information resulting from accumulated knowledge must be made available and understood by those who will be responsible for its application.

Some details of life expectancy of selected components are given at Appendix B.

An experienced maintenance manager should be co-opted to the design team at an early stage in order that as many maintenance problems as possible may be designed out in these early stages, and in particular to ensure that proper access and facilities for repair are included.

Other areas requiring particular attention include:

● adequate allowance for thermal movement;

● correct detailing of damp courses;

● detailing of joints between materials and finishes;

● falls to flat roofs;

● general attention to the effects of wind and rain, particularly in exposed areas;

● consideration of the conditions that bring about excess condensation and mould growth;

There is a need for better site supervision and quality control, as bad workmanship and inaccurate detailing make a very significant contribution to subsequent maintenance costs.

There is less scope for increasing productivity in maintenance work than in new work, because in addition to fixing items, craftsmen must spend time in diagnosing faults, tracing services and removing defective items. It is more labour intensive than new works, to which modern technology and mechanical aids can be applied more readily.

Ideally, building maintenance operatives should be carefully selected craftsmen who are trained in diagnostic and maintenance techniques, and who have understanding of the cause of failure, and the way to avoid it in the future.

It is necessary to specify materials correctly in the refurbishment and repair of older buildings and care must be taken in considering the effect of blending modern materials with traditional construction.

Research into the cause of failure should, in time, provide ways in which maintenance can be reduced, particularly if manufacturers and designers use the information as a basis for further research and development of materials and techniques. Designers must also learn of faults identified from the feedback of information, so that they are not repeated in the future.

Those concerned with maintenance should have a wide knowledge and understanding of the buildings for which they are responsible. One way of achieving this is for the design team to produce maintenance manuals as part of its commission. (See Section in Chapter 5 on 'Maintenance Manuals'). They should ensure that speedy decisions are made and problems identified in order that the remedial works may be undertaken before further deterioration occurs.

Chapter 2
Maintenance Defined

DEFINITION AND SCOPE

British Standard BS3811* defines 'maintenance' as:

> 'Work undertaken in order to keep or restore every facility, ie every part
> of a site, building and contents to an acceptable standard.'

However, this definition is more generally seen in an engineering context
and The Chartered Institute of Building supports the following definition:

> 'Building maintenance is work undertaken to keep, restore or improve
> every facility, ie every part of a building, its services and surrounds to an
> agreed standard, determined by the balance between need and available
> resources'.

Maintenance strategy is aimed at maximising the fixed assets of a company
or organisation, but even so, maintenance budgets are very vulnerable to
changes in the economic climate. In a recession they are one of the first to
be singled out for reductions and even in good times there is a general
reluctance to spend sufficient sums. Normally, statutory requirements are
met, although these tend towards minimum rather than optimum standards.

In an extreme case, where inadequate sums are spent on maintenance, the
'value of the facility' as a commercial asset, may be converted into a liability
and in such a case it can be argued that the maintenance policy has failed.

Maintenance spending is normally governed by annual budgets and this has
given rise to the traditional view that maintenance is a short term service
operation financed from current account spending. In fact, buildings have a
long life span and maintenance should be viewed in this larger time frame.

Maintenance is a programmed transformation of a building's fabric and
service, reflecting changes in patterns of use and technology. In the early
years of a building's life, work will concentrate on keeping it near to its
original state. Later, the introduction of a new technology, or the changing
needs of the people who live or work there, may call for a refurbishment
programme. This may entail extensive upgrading, the provision of new
capital projects and may later extend to rehabilitation or restoration work.
All of these developments can fall within the orbit of maintenance management and underline the need for forward thinking and planning.

---

*BRITISH STANDARDS INSTITUTION. (1974) Glossary of maintenance terms in terotech-
nology. BS3811.

CLASSIFICATION OF MAINTENANCE**

Maintenance is usually of three types:

- preventive maintenance;
- routine maintenance;
- emergency maintenance.

Preventive maintenance, also called **planned maintenance,** consists of taking corrective or preventive action in order to avoid expected or avoidable failures. A well thought out plan can save considerable sums of money, although the need for cost effectiveness must be borne in mind. For example, monitoring schemes may be more costly than the periodic replacement of items with a predictable service life.

Routine maintenance is the routine or day-to-day repair or replacement of defective items.

Emergency maintenance is the speedy action needed to rectify failures. These can stem from failures in planned maintenance or from natural or other causes, eg. storm, flood, health reasons, accidents and security. They are invariably disruptive and expensive and highlight the need for an effective planned programme of maintenance.

**Note: There is inevitably some degree of overlap between the types of maintenance and 'corrective' maintenance is sometimes designated as a separate entity. Although not specified separately, 'corrective maintenance' is considered within the categories given above.

Chapter 3
Organisations

The national annual expenditure on the maintenance of buildings, engineering and other services in this field runs into many billions of pounds. The effect on the economy, and on the building industry, is enormous, so it is pertinent to ensure that the maximum efficiency is obtained by estate managers from work undertaken by contractors, direct labour organisations and specialists. Value for money and good service must be two major objectives in setting up a maintenance organisation.

Whereas it is true to say that this country has a long history of good maintenance and probably leads the world in this subject, there is still room for considerable improvement.

GENERAL RESPONSIBILITIES

- Preparing estimates, making inspections, preparing drawings, specifications and contract documents for obtaining tenders and quotations for all maintenance work and repairs. Ordering, supervising work and certifying accounts for payment; managing all direct labour operatives employed in the organisation. Financial management, materials and stores management, plant and transport management. Work planning, programming and budgeting of maintenance and servicing.

- Negotiation for, ordering and checking utility services (eg. gas, electricity, heating, fuel efficiency, water, drainage and sewerage) and their efficient management as they affect the client.

- As above for minor works, renewals and improvements.

- Providing a consultancy and advisory service to the building occupier. Employment of an outside consultancy specialist to the client as required.

- Ancillary services–porters, cleaners, fire precautions, security, waste disposal, gardens and grounds.

TYPE OF ORGANISATION

The type of organisation to be set up will depend upon the function of the parent body and upon the responsibilities allocated to the maintenance organisation.

As a guide to the building owner on the type of organisation required to meet the needs of his estate, six main elements have been listed in Figure 1 for four typical building owners. Note: It is not possible to define a single maintenance organisation suitable for all types of maintenance work.

The six main elements are:

- the maintenance and operational workload. This can be assessed and includes in the calculation other service standards, cyclical maintenance, preventive and planned maintenance;

- the geographical area over which the workload is spread;

- work other than maintenance for which the organisation will be responsible;
- the building owner's maintenance policy for different groups or types of buildings;
- the builder owner's or user's maintenance requirements;
- professional staff to meet the above.

PRINCIPLES OF ORGANISATION

To illustrate the major principles, four typical building owners are used as examples. Whilst they may be considered to have large estates, most of the points raised apply to all building owners.

The estates

Each owner has an estate comprising at least $400,000^3$ gross floor area of buildings.

Building Owner A

A national company with 1,200 retail outlets located in major towns and cities, a large headquarters, office, factory and warehouse and research laboratories.

Building Owner B

A large county authority with mixed properties, comprising police stations, courts, houses, clinics, fire and ambulance stations, schools, colleges and offices.

Building Owner C

An industrial company with a site area of 500 acres containing a prestige office block, research laboratories and several production lines and storage buildings.

Building Owner D

An institution on a site of about 300 acres, eg. university or college, with buildings occupied 24 hours each day.

The workloads

The management structure required to meet the needs of the four typical building owners can now be assessed:

Building Owner A (National retailing company)

Brief assessment of workload

The maintenance requirements of the shops are governed by time and business restraints. Major repairs to roofs, floors and fabric are carried out after inspections. Redecorations are programmed. Maintenance at the factory, stores and research laboratories is of particular importance, since a rapid response is required for repairs to electrical equipment and the main power sources.

		Owner A	Owner B	Owner C	Owner D
1	Maintenance work load	a 1,200 retail shops b Factory and warehouse c Research laboratories d Headquarters' offices	a Police, fire and ambulance stations b Houses c Courts d Schools and clinics e Colleges f Offices	a Very large factories and stores b Research buildings c Several office blocks	a Academic and research buildings b Administrative and communal c Residential d Recreational
i	Age of buildings	Varying from a hundred years to modern structures	Varying from several hundred years old to modern structures	Varying from 150 years old to modern structures	Buildings 80% less than 20 years old, few over 100 years
2	Geographical area of workload	a National - in all major towns; b ( ); c ( ) on one site covering about; d) 300 acres	Covering several hundred square miles	Covering approximately 500 acres	a, b, c Within a radius of 3 miles covering 300 acres; d Within 8 miles
3	Work other than maintenance				
i	Expansion - buying and selling properties	Small amount, two or three	Buying and selling all the time	Little	Small number
ii	Designing new buildings	Two or three per annum	Continuous programme	Intermittent	Small number
iii	Alterations and upgrading existing properties	Continually upgrading all shops once every 15 years; represents 80 per annum	Alterations; infrequent upgrading	Alterations, some upgrading	Large number all the time
4	Owner's maintenance policy	a Policy of minimum maintenance, using maintenance free materials and fittings to maintain public image. Short term upgradings b Minimum of maintenance (within legal requirements) consistent with profits and funds available c & d Prestige buildings, high standard of maintenance and finishes	a-f Maintained to a standard consistent with type of building, age and funds available	a Standard to comply with legislation and funds available from profits b & c Prestige buildings, high standard of maintenance and finishes	a Good standard maintained b to keep to original c facility provided d Maintained to standard consistent with funds available
5	Building owner's maintenance requirements. These depend upon the prime function of the buildings in each group. The following need an answer for each group.				
i	Hours in use each day?	a 0800 to 1800; b & c Part continuous; d 0800 to 1800	a & b continuous; others mainly 0800 to 1800	a & b Part continuous, part 0700 to 1800; c 0800 to 1800	a from 0700 to 1800, partial use until 2200; b 0700 to 2200, partial use until 0200 next day; c 24 hrs; d 0700 to 2300.
ii	Days in use each week?	a 6; b & c 7; d 5	a & b 7; others 5	a & b Part 7 part 5; c 5	a 5 (partial use for 7); b 6; c 7; d 7
iii	Is maintenance restricted in any way?	a Yes; b & c Parts no, food parts yes; d No	a, b, d No; Others some restrictions	No	Only in hazardous zones
iv	Is there a major engineering service of motive power?	a Electricity; b & c Steam & Electricity; d Electricity	None	a & b Steam, electricity & gas processes; c Gas and electricity	Yes, steam and electricity
v	Will iv need an emergency service?	a No; b & c Yes; d Yes (computer)	No	Yes	Yes
vi	Is there domiciliary staff?	During working hours a Manager; b & c Engineers; d Yes	In some cases yes, in schools and colleges	No, but engineers during working hours	No
vii	Are craft skills required?	a No; b, c, d; fitters and electricians	Carpenters, plumbers, electricians	Fitters and electricians	Fitters and electricians
6	Professional staff required				
i	Estate management & valuation	Yes, for expansion	Yes	Yes, during expansion	Yes, frequently
ii	Design teams	Yes, for new works	Yes	Yes, occasionally	Yes, for new works
iii	Maintenance management	Yes, for alterations, upgradings and maintenance	Yes	Yes, or works engineer	Yes, alterations, upgradings and maintenance

Figure 1. Four typical building owners

9

Workload in detail

(a) Building fabric repairs and redecorations

The heaviest workload is at the headquarters and the factory site, so this is where the maintenance manager is located. To reduce unproductive travelling time to a minimum, the workload is split up into districts, with an assistant maintenance manager in charge of the buildings within that district. Consequently, there are four or five assistant maintenance managers.

An alternative solution is for maintenance management at headquarters to use local, well established building contractors to prepare annual or biennial full inspection reports with estimates for any required work which later form the basis of a contract. In between annual reports the contractor deals with all day-to-day repairs. (Term contracts by competition for a 12 months period is possibly the best way to test the market place price).

(b) Electrical engineering

The only amount of work sufficient to justify a full time qualified engineer is at the HQ and factory site. The other single shops are looked after in a slightly different way. Small tasks such as changing individual lamps and repairing fuses are done by a handyman under the jurisdiction of a shop manager. For larger jobs or specialist work the manager is empowered to use an established local electrical contractor, or such a contractor through the recognised building main contractor. Specialist items, such as refrigeration and deep freeze equipment and outside lighting are carried out by a sub-contractor under a 'service agreement'.

(c) Mechanical, heating and ventilating services

Again, only the HQ and factory site justify a full time engineer, who arranges for the servicing of mechanical plant to be carried out by a local firm.

Building Owner B (County authority)

Brief assessment of the workload

The authority has a variety of properties, many being relatively small units (500^2 gross floor area or less), except for the schools and colleges etc. County offices and their satellite offices are comparatively larger.

The number of day-to-day minor building maintenance items is small for the many smaller units and is more concentrated at the main county offices and other larger units. Major repairs, especially to roofs, hardstandings and playgrounds, fences and boundaries require time for inspections and specification writing, as does the redecoration programme. There is also alteration work to the considered.

Electrical repairs follow a similar pattern, except that many of the older properties may need re-wiring or upgrading. The load is concentrated in schools, colleges and the larger offices. Additional work is associated with building alteration.

For mechanical or heating and ventilation services, there is also a similar pattern in maintenance demand but at a lower level. There is some air conditioning, including the computer facilities. Building alterations impose an additional workload.

Workload in detail

(a) Building fabric repairs and redecorations

The workload is sufficient to justify the appointment of a maintenance manager. The workload of the county is divided up into divisions or areas.

As the administrative area is large, there are area offices in the main towns where one or more area maintenance managers operate, with appropriate secretarial and clerical assistance. The objective of the system is to reduce non-productive travelling time to a minimum.

Qualified maintenance managers could be used to deal with the professional aspect of the work, such as surveys, drawings and specifications and clerks of works for supervision.

Whether this method is employed depends to a great extent upon the confidence that exists between the building owner's organisation and the contractor and his organisation. If it is found necessary to have daily site visits, or something approaching that frequency, then the employment of a clerk of works might be the most economical solution.

There are many hundreds of buildings forming this estate and in the majority of cases there is a person domiciled who can deal with breakdown situations and the smaller day-to-day items. For example, caretakers or head teachers at schools and station sergeants at police stations and courts. (If the workload justifies the full time employment of direct labour staff, this could be provided on a mobile basis).

(b) Electrical engineering

With the increase in sophistication of modern buildings and in the standards of facilities provided, there is a need for qualified electrical engineers.

(c) Mechanical, heating and ventilating services

There is a smaller workload for the mechanical engineer than for the electrical and, as the training of both covers some ground of the others discipline, they carry out both types of work.

Building Owner C (Industrial concern)

Brief assessment of the workload

In addition to the building and engineering services normally associated with the buildings, there is a very large amount of production engineering equipment. The policy is to integrate the maintenance responsibility under one control, thus the building repairs take second or third place to electrical and mechanical repairs.

The workload is so concentrated that the maintenance team is split up with a small unit for each of the larger production units. There is a central workshop and stores and maintenance administration.

It is vital in such a situation that alterations and additions are integrated with maintenance, so that interruptions of production are reduced to a minimum. Since there is a continuous process in operation and long run research projects, maintenance to cover these situations is provided by a seven day shift working of emergency gangs.

Workload in detail

(a) Building fabric repairs and redecorations

The majority of maintenance results from constant use and rough handling. Doors and all moving parts need good quality materials and fixings. Floors and drains need special attention, as do roofs and gutters. The facility to carry out redecorations is restricted and virtually all work needs programming in advance.

(b) Electrical engineering

Electrical repairs need a rapid response, as electricity is one of the main power sources. Quick diagnosis of faults is the key, together with good spares and standardisation. Planned programmes and replacements with bench rewinds should be borne in mind. Safety precautions must have a very high priority. Mechanical robots require special attention. Computers and micro electronics require specialist contracts.

(c) Mechanical, heating and ventilating services

In these areas the engineer is the maintenance team leader, particularly where the motive power source is steam. There is a first class schedule of inspections and periodic maintenance, with well organised emergency procedures and teams. Good records of plant and parts, together with good stocks of spares are available. Mechanical robots require special attention.

Building Owner D (Institution)

Brief assessment of the workload

In addition to buildings and engineering services, glass-houses are included for botanical and genetic teaching and research, as well as animal breeding accommodation. There are several computer installations and long run experiments which take place all the year round.

Workload in detail

(a) Building fabric repairs and redecorations

The main workload is the annual redecoration programme and all other major works are centred on this. Repairs to roofing, roads, car parks and hardstandings are a major annual expenditure, as are floor repairs. Alterations provide a constant source of work.

(b) Electrical engineering

Lift maintenance, lamp changing, lamp cleaning, fire alarms, emergency lighting and clock systems are major items, as is the upgrading of lighting and rewiring. Alterations provide a constant flow of work. Computers, micro-electronics and other specialist equipment receive specialist servicing.

(c) Mechanical, heating and ventilating services

The major item is the efficient running of the boiler plant and the distribution of essential services, heating, steam, domestic hot water, gas, cold water and fire mains. Maintenance to controlled environmental plant and fume extraction systems create a big workload. Alterations involving these items are complicated.

Requirements of professional staff

All four building owners require (in-house or external):

(a) professional advice on estate potential, utilisation and land transaction;

(b) design team(s) for new buildings;

(c) maintenance management for repairs, maintenance and operations;

(d) maintenance management for alterations and upgrading of buildings.

Figure 2 shows various ways of dealing with the above four elements. For maintenance and alterations to existing properties the first method for all four building owners is to use directly employed maintenance staff. In all cases there are reasons why the control of drawings and details of assets should be under strict control. This control is more easily maintained by using directly employed staff, with documentation under some form of security control. In addition, lines of communication between the buildings owner's main department and the maintenance organisation will be more effective than by either using external consultants or contractors.

Other building owners with a smaller workload than that under consideration may nevertheless have a sufficient workload to justify at least one qualified and experienced maintenance manager to run the department efficiently and economically.

Function	Method One	Method Two	Method Three
Estate potential utilisation and land transactions	Building Owners A & B have a workload to justify directly employed staff	Building Owners C & D need external consultants	
Design of new buildings	Building Owners A & B require staff architects an services engineers either under one person or independent	As Method One assisted in peak workload times by external consultant. Building owners C & D need a retained architectural practice and/or services engineers	Contractors
Maintenance and operation of existing properties	All four have a workload justifying directly employed management staff	External consultants	Contractors
Alterations and upgrading existing buildings	All require directly employed management staff	External consultants	Contractors

Figure 2. Alternative methods of determining professional staff requirements

MAINTENANCE MANAGEMENT

Management structure

Building Owner A

With such a large number of small work units, Building Owner A will not require a fully integrated management structure, but, for his headquarters and factory site, may well have all five sections as shown in Figure 3 integrated under one company director.

Building Owner B

It has been unusual to find a fully integrated management structure, on the lines of Figure 3, in local government organisations. Firstly, the buildings 'belong' to several policy making committees and finance for their upkeep is dealt with by each; consequently different standards may be applied. Management of the building may be by the official in the 'owning' department with the professional and technical staff in other independent departments. However, this is now changing with independent building works departments being set up.

Corporate management used in many authorities has achieved varying levels of success but the departmental manager generally has to accept responsibility when performances are unacceptable to the authority.

The Technical Department, eg. architects, engineers, surveyors, etc, is usually controlled by a director of technical services.

Finance and accounting are the responsibility of the Treasurer and Valuation Officer in another department.

Building Owner C

This owner could employ a highly integrated management structure, under one of the company directors, with the works engineer of an individual factory unit as the co-ordinator for sections 1 to 3. Headquarters would then integrate sections 4 and 5.

Building Owner D

As the workload for this client is concentrated over a relatively small geographical area, it provides the best situation for a fully integrated management structure, both horizontally and vertically and can include a certain amount of direct labour.

Maintenance requirements

Little has been mentioned about this element which sets out the building owner's requirements to be provided by the maintenance organisation. The details on which decisions have to be made are based on the degree of response to the request for repair and how it can be met by the maintenance organisation.

Building Owner A (National retailing company)

As the prime function of the 1200 shops is to sell goods, it is imperative that each counter unit is available during the hours of opening. Most emergencies can be handled by the shop manager, or by the local contractor with whom he may have a contract.

At this building owner's HQ and factory site there is a work force of direct labour to meet maintenance requirements for motive power and air

Figure 3. Integrated management structure

				SECTION 1 BUILDING	SECTION 2 ELECTRICAL	SECTION 3 MECHANICAL (H & V)	SECTION 4 PLANNING METHOD STUDY	SECTION 5 ADMINISTRATION	SECTION 6 ANCILLARY SERVICES	SECTION 7 GROUNDS
MAINTENANCE ORGANISATION	Governing body	Council	or	Board of Directors						
	Type of owner	Building owners B and D	or	Building owners A and C						
	Responsible officer	Chief officer(s)	or	Director(s)						
	Property management	Estate surveyor	or	Consultant agent						
	New building design	Design team directly employed	or	Consultant architect and engineers						
	Maintenance { in existing / Alterations { buildings	Maintenance manager								
	Main head of Expenditure			Senior surveyor	Senior engineer	Senior engineer	Estimators	Secretarial	Porters and cleaners	
	Professional and technical staff			Surveyors Regional District Area	Engineers Regional District Area	Engineers Regional District Area		Clerical Wages Invoices Stores purchase Costing	Security-day Security-night Fire precautions	
	Supervision			Clerks of works	Clerks of works	Clerks of works				
EXECUTION OF WORK	Operational			Foreman/Foremen — Carpenters Bricklayers Plumbers Painters Labourers Drivers	Foreman/Foremen — Electricians Lift mechanics Bench fitters Lamp changers Lamp cleaners Mates Drivers	Foreman/Foremen — Pipefitters/Welders Bench fitters Mates Drivers				

15

conditioning plant for the computer. Apart from the small emergency service provision, the majority of work can be carried out by contract or a mobile direct labour force.

Building Owner B (County authority)

The answers to element 5 (Figure 1), subsections iv and v are negative and thus there is little need for quick response. Most buildings have someone living on site who can deal with emergencies, and a telephone call to the assistant maintenance manager will invoke quick response from the local contractor or a mobile direct labour force.

There would appear to be little need for a large direct labour force except perhaps, in the town where the head offices are situated. Even there, with contractors close at hand, it is possible to have a term contract to provide a good service. The majority, if not all of the work, can be carried out by contract.

With the reorganisation of local government, which took place in April 1974, responsibility for certain functions has changed and buildings which are formerly under one authority have been transferred to another larger authority.

There are now greater concentrated workloads and in particular large housing estates. Where these large workloads exist a nucleus of direct labour working from a depot can provide quick response to emergency calls with mobile teams for dispersed estates and buildings.

Peaks in the workload and large seasonal jobs can be put out to contract on a term contract or a separate contract for larger projects.

Building Owner C (Industrial concern)

On reading the requirements set down in element 5 (Figure 1) it is obvious that this owner needs a continuous quick response to his request for repair. This can best be provided by men being on hand, close to the positions of high workload. This is a classic case for using directly employed labour who should have available modern mechanical tools, transport and stores, to provide a quick and efficient response. Close control of such staff is essential for good performance.

It must be borne in mind that in addition to the repair work itself, the profitability of the company may be at risk in any work stoppage of production, or walkout by work people, if working conditions deteriorate. Apart from this emergency service provision, it is possible to organise and plan the remainder of the maintenance work to be done by contract if required.

Building Owner D (Institution)

As there is a continuously operating main power source and a requirement for an emergency service every day of the year (animal breeding facilities and air conditioning), service can best be provided by directly employed labour acquainted with the work proceeding within the buildings and aware of any hazards of special equipment.

People occupying these buildings demand a high standard and expect a fairly rapid response to their requests for repairs. The direct labour sections, therefore, should be sufficiently well manned and organised to provide a speedy response, consistent with efficient and economical standards.

All specialist items of repair are put out on service contracts and all other major items out to competitive, selective or negotiated contract.

In the preceding paragraphs the requirements of four similarly sized estates with different types of owners have been outlined. From these it is possible to derive various management structures to meet individual requirements.

Qualities and qualifications of maintenance staff

Whether an organisation uses directly employed labour or outside contractors for its maintenance functions, its own staff controlling or overseeing the operations should have the ability to perform effectively. This is not always so in reality.

The categories and number of people employed on maintenance activities will obviously vary from organisation to organisation to organisation.

The maintenance or facilities manager should ideally be a professional with CIOB qualifications. Some years of actual experience in building maintenance and management are also of great advantage.

For smaller units, an HNC or BTEC qualification in an appropriate discipline could be sufficient for effective management. It cannot be emphasised too much that the manager should have adequate technical knowledge to control correctly the activities of his subordinates.

In general, maintenance operatives tend to be of a mature age. It is hoped that courses containing a maintenance element will encourage younger men to specialise in this subject.

Craftsmen should hold the requisite City and Guilds or other craft certificates. In addition they should enjoy working either on their own or in small groups and still maintain responsibility to their employer and client.

The acquisition of and the will to use multi-skills maintenance is to be encouraged.

In the larger organisations it is necessary to interpose supervisory or organisational personnel between the manager and the operatives. These people should have qualifications and experience in their particular sphere. They should, above all, be able to communicate both upwards and downwards.

Inevitably, there will be functions of a recording and budgetary nature, and it is assumed that appropriate persons will be selected.

Finally, the importance of the maintenance function as a whole suggests that remuneration should be such as to encourage all concerned to have high production rates together with responsibility for their actions.

Chapter 4
The Maintenance Strategy

,UIREMENTS

- What is required to be done?
- Which items are the most important?
- What are the legal requirements?
- When can the work be done without loss of production or facilities or service?
- How often should inspections, surveys, tests and works be carried out?
- Where is the money coming from?

These are some of the questions to which the maintenance manager must find the answers in order to formulate his maintenance strategy and produce his maintenance plan or programme.

CONDITION SURVEY AND MAINTENANCE PLAN

Before a maintenance programme is prepared, a complete inspection of the estate should be carried out. This will include physical close inspection of the structure and its external cladding, all internal surfaces and fittings, the roof space and all services together with the surrounding estate grounds and external facilities. Electrical installations, lightning conductors and any other wired systems should be tested.

The notes and comments are set out in a comprehensive report supplemented by drawings and photographs. Recommendations on works required and their priority are included. This document is termed a 'Condition Survey' and will prove invaluable in compiling a maintenance programme.

If resources permit, it is recommended that a condition survey is carried out on each estate or property every five years.

Providing that external redecoration can be carried out within twelve months of the production of the condition survey report the repairs to items such as gutters, downpipes and external joinery can be extracted from the survey report and embodied in the specification with the decoration so that they may be completed immediately prior to repainting.

The work necessary as indicated by a condition survey can be programmed according to its priority and executed over a period of the next three or so years.

Some elements of properties such as flat roofs, gutters and downpipes require more frequent inspection, preferably on an annual basis. Inspection of a flat roof annually will indicate when the covering will need to be renewed in whole or in part. Acting promptly on this information may well avoid the expense of internal redecoration and possible loss of facilities and income resulting from flooded rooms or assembly halls or even more serious problems to the structure arising from the development of dry rot or similar infections.

Having been costed and programmed the maintenance plan should be supported by a firm recommendation and presented to the board, committee, trustees or senior management as the case may be. The presentation must include the order of priority of tasks so that decisions are made according to need and do not reflect solely the financial allocation in the current or forthcoming year.

Once the plan has been approved in whole or in part it forms the basic programme of work, bearing in mind that some flexibility will still be required as priorities and circumstances may change. As changes are disruptive and may delay the whole programme they should be kept to the minimum.

LEGAL RESPONSIBILITIES

Legal responsibilities affecting maintenance are very wide. They cover not only the fabric of the property and its equipment but also the owner, the occupier, classes of vistors and the general public. The relevant Acts of Parliament are designed to safeguard people who may need to enter or use the property.

The Health and Safety at Work etc Act 1974 must be complied with strictly in the execution of all building maintenance work. When properties are occupied or in use when maintenance is carried out the maintenance manager has a shared responsibility with the contractor for the safety of all users and vistors. Where work is carried out to asbestos in or on buildings the Asbestos Regulations must be followed.

Public Health Notices issued by the Environmental Services Officer of the local authority must receive attention and action within the specified period.

The Landlord and Tenant Act, Town and Country Planning Act and the Building Regulations (1985) are designed to safeguard property as an asset, an amenity or at least to prevent it being a nuisance to others.

In England and Wales the Building Control Officer must be notified, or an Approved Inspector employed in respect of any 'Building Work' as defined in Regulation 3 of the Building Regulations 1985.

The current definition of 'building work' differs from previous legislation and includes material alterations to the requirements of the Regulations concerning means of escape from fire, internal fire spread, controlled services and fittings, as well as the structure of the building, the installation of insulating material and works involving underpinning.*

Attention will need to be given to:

notices received from a local authority under Sections 76-80 of the Building Act 1984, concerning defective or dangerous buildings and demolition work;

obtaining any necessary hoarding or scaffold permits from the Highway Authority for works affecting or abutting streets;

---

*Further information can be found in the *Manual to the Building Regulations 1985*, published by HMSO.

observing the requirements of the Noise and Control of Pollution Act where noise or dust etc is likely to cause a nuisance;

notifying the local fire authority for any material modifications or works which may possibly affect any conditions contained in a Fire Certificate issued under the Fire Precautions Act, 1971.

The maintenance manager has a responsibility for ensuring that the legal requirements are met and that appropriate action is taken and documented. This may entail seeking expert legal, professional or scientific advice but this in itself does not remove the burden of responsibility from the maintenance manager.

WHEN CAN THE WORK BE CARRIED OUT?

Figure 4 provides examples of the types of property for which maintenance managers may have responsibility, together with an indication when major maintenance and redecoration to the internal areas may be executed. External maintenance such as joinery repairs, repointing and redecoration can usually be carried out in normal working hours. Reroofing may also be included in this category, with the exception of flat roof coverings to educational establishments, which are preferably executed during holiday periods recesses. Small internal repairs can usually be carried out during normal working hours with the exception of operating theatres, other specialist hospital facilities, consulting rooms, catering establishments and such like.

It is most important that all work (except real emergency work) is preplanned and the timing duration agreed with the manager or senior person responsible for the premises. All alterations to the agreed works and timetable should also be the subject of consultation and further agreement.

When work is executed over several nights or weekends and the building is being used in the meantime it is essential that particular care is given to:

the removal of materials, plant and rubbish from all areas to be occupied;

leaving all areas clean and tidy;

replacing all covers which may have been removed from ducts, manholes, cleaning eyes etc;

ensuring that gas, electric, telephone, security systems, water, waste and sewerage systems are operational and safe;

ensuring the safety of occupants, users and visitors by providing adequate protection and warning notices of any hazards which cannot be removed at the end of a night or weekend working period;

complying in all respects with the Health and Safety at Work etc Act 1974 and the Asbestos Regulations.

ORGANISATION	EXAMPLES OF PROPERTIES	ACCESS FOR MAINTENANCE
Government	Public offices, workshops	Public holidays, evenings and weekends
Public utility	Offices, shops	Public holidays, evenings and weekends
Municipal authority	Housing estates Offices Leisure facilities	Normal working hours Public holidays, evenings and weekends Out of season, nights and special-closures
Educational establishment	Lecture halls, classrooms Living accommodation	During recesses: minor works evenings and Normal working hours
Hospitals, public or private	Operating theatres Wards Living accommodation	Special closures required Special closure except for small emergency works Normal working hours
Industrial company	Factories	Holiday periods, evenings and weekends
Commercial company	Offices	Public holidays, evenings and weekends
Hotel	All areas	Out of season closedown or special closure

Figure 4. When internal maintenance work can be carried out

MAINTENANCE STANDARDS

There are two main divisions of maintenance standards:

● Quality standards

● Service standards

Quality Standards

Quality standards depend on the products used, the skill of the operatives, the standard of supervision during the work and the approval (or rejection) upon inspection of the completed work by the maintenance manager.

Quality assurance (QA) is a growing movement in building and other industries. Companies participating in the scheme give guarantees in respect of their products, their craftsmanship or both.

Service Standards

It is important to quantify service standards for maintenance in terms of time periods or response times in which the work is to be completed. From this, programmes of work can be prepared and performance measured against the standards. Service standards are, therefore, essential as a basis for calculating the workload together with labour requirements (for in-house staff or contractors) and the financial resources needed.

The activities included in the maintenance service standards should cover for example, external painting, internal redecorations, emergency repairs, trade repairs, ie. bricklaying, carpentry, plumbing, glazing, electrical, space heating, boilers, horticultural work and cleaning (see Appendix B. Maintenance service standards).

Time periods should be agreed for servicing the following major items:

gas and electrical installations;

ventilation and air conditioning systems;

drains;

hot and cold water services;

scale reducers;

petrol interceptors etc;

servicing of plant, machinery, lifts, escalators, fire fighting equipment, dry risers, fire doors, escape routes.

It is also important that standards are included to cover legislation, including the Health and Safety at Work Act, Defective Premises Act, Housing Acts, Factories Acts, etc. Inspections should be fully documented and bear the signatures of the officers carrying out the inspection.

SOURCES OF FINANCE

Internal

- current cash flow–generated from sales and service;
- savings and investments–money previously set aside from the profits on savings and investments;
- property–proceeds of the sale or lease of properties or the use of property as a security for a loan;
- reserves–monies set aside to meet the cost of specific projects;
- rates–monies raised from the rate payers.

External

- shareholders–issuing of further company shares or rights issues;
- banks and other financial institutions–short or long term loans or overdrafts;
- credit–negotiated special credit terms with suppliers;
- leasing arrangements–plant and equipment may be leased or sold to raise cash;
- tax allowances/concessions/grants–tax advantages and grants are available for a wide range of property investment and repair in both the private and public sector;
- local authority loans;
- government funds.

VALUE FOR MONEY

Local government

If a local authority employs its own work force, then the use of a work study based incentive bonus scheme, properly managed, can result in high productivity levels from operatives.

Time values for elements of maintenance work can readily be obtained by work study, and daily allowances made for travelling, gaining entry to work place, collecting tools and packing away, sharpening tools, making out time sheets, washing hands, fatigue, etc. The total standard hours produced provide the basis for calculating productivity payments as well as information for management on performance.

Motivation of other personnel is probably best achieved by management by objectives (MBO), geared to the performance of operatives and supervisors. The total cost per productive standard hour (PSH) achieved by the work force can be compared with contractors' costs on a work sampling basis, where the total cost per PSH in each case is multiplied by the total PSH's in the sample. It is important that this comparison be made regularly to monitor cost performance. For comparison purposes, all overheads, establishment charges and return on capital employed must be included in the cost of the in-house work force. To the contractor's cost must be added a charge to cover supervision, documentation, checking and payment of accounts.

Maintenance work by contractors can be effected either by contract by competitive tender or quotation for individuals jobs. In this case the job is usually of some size, requiring at least drawings and a specification.

Alternatively, term contracts may be used for day-to-day repairs, and are generally let for a two or three year period. A schedule of rates is obtained from a detailed repairs schedule or specification by separate trades. This will give the lowest market place price for day-to-day repairs.

Contractors

To obtain a realistic quotation from a specialist or general contractor, it is most important to prepare a compact, umambiguous and detailed specification, together with drawings where necessary, for all the work or services required. Inadequate specifications generally lead to high estimates. It is essential that contractors quoting for work must receive identical documentation.

Other points to remember when letting a job to a contractor are:

- orders for the work must be clear;
- payment for satisfactory work should be prompt;
- additional work required, or variations on contract, must be thoroughly documented and prices obtained before variation orders are issued;
- the contract must not be overspent without written authority from the client;
- supervision of the work must be adequate to ensure satisfactory quality control and that progress complies with the approved work programme;
- the client must ensure that the contractor has adequate access to the work place and that unavoidable delays are minimised.

FORMS OF CONTRACT

The Joint Contracts Tribunal currently issues three different forms of contract which are suitable for most maintenance works subject to competitive tendering. They are:

The Agreement for Minor Works 1980 Edition with current amendments;

The Intermediate Form of Building Contract (IFC 84) with current amendments;

The Standard Form of Building Contract 1980 Edition (JCT80) with current revisions. There are two editions, one for local authorities and another for all other employers. Both editions have variations for use with or without bills of quantity.

There are no hard and fast demarcation lines to determine when a particular form should be used but the following will provide useful guidance.

Minor Works Agreement

This is suitable for works for which a lump sum offer is being obtained, based on drawings and/or specifications and/or schedules but without detailed measurement. The form is generally suitable for contracts on the above basis up to a value of £8,000 at 1989 prices.

IFC 84

The cover to this form of contract bears the words 'Form of Contract for works of simple content'. Despite this description it is suitable for works of greater value than the limit suggested for the Minor Works Agreement. It does provide for the use of bills of quantity where required and for the use of sub-contractors named by the employer. It also provides for the adjustment fluctuations in the prices of labour and material valued added and other taxes. It is suitable for works which are not of a complex nature up to approximately £350,000 at 1989 prices and possibly for very simple works such as external redecoration and associated minor repairs up to £500,000.

JCT80

This form of contract is suitable for works of a complex nature or having a value in excess of the limits suggested for the IFC 84.

Other forms of contract

The Joint Contracts Tribunal also issues forms of contract for tendering on a management fee basis and a design and build basis.

The Association of Consultant Architects issues a form of contract for use with major projects.

The Architects and Surveyors Institute publishes two forms of contract, one suitable for minor works and the other for major works.

DAYWORK

When daywork is considered to be a suitable basis for the valuation and payment of maintenance work it should be contractually linked to the *Prime cost of daywork charges for building work* published by the Royal Institution of Chartered Surveyors.

CALL OUT CHARGES

Many specialist contractors quote for emergency works on the basis of a call out charge plus a rate per hour or part of an hour for the time spent on site by each operative. Both call out charge and hourly rate generally increase for evening and weekend responses. Its is expedient to negotiate charges annually with specialist contractors likely to be employed on this basis. In this way reductions can be obtained and the rates are known before a call out occurs.

ANNUAL OR REGULAR SUMS

There are maintenance operations which will be required at regular intervals, some annually such as servicing to boilers and other gas applicances and others more frequently such as window-cleaning. Competitive quotations can be sought in advance and if contracts are placed on say a three year term for gas services or an annual term for window-cleaning, it is possible to negotiate discounted rates.

Chapter 5

Management of Maintenance

MAINTENANCE MANAGEMENT AS A DISTINCT FUNCTION

Nearly one half of the industry's effort is involved in maintenance and it has already been indicated that this represents an enormous investment. Traditionally, greater recognition has always been given to new work, with maintenance being considered of much lesser importance. However, there has always been a need for maintenace and this, particularly over recent years, has been emphasised by escalating costs. Building owners are no longer prepared to accept buildings which are costly to run and to maintain. The need to calculate life cycle costs is becoming an essential consideration.

To this end, many organisations have appointed their own maintenance managers and this has shown a significant improvement in achieving value for money.

Those involved in new work must fully understand the technology and special needs related to that particular project. This is also a requirement for those involved in its subsequent maintenance.

This has led to the identification of specific and separate activities. Maintenance management is now recognised as a distinct function in its own right.

PRINCIPAL CONSIDERATIONS
Manpower

Craftsmen trained on new work tend to drift into maintenance when they can no longer compete with younger men. In some respects the experience gained on new work may be very useful, but there are disadvantages because they may be set in their attitudes and in the use of working methods not specifically suited to maintenance operations. Very few have been trained specifically in maintenance, although, with the increase in local authority and institutional direct labour departments, this situation is changing. The development of progressive career structures and the implementation of incentive schemes within maintenance is attracting younger operatives. There is a noticeable and welcome trend for new project managers to be chosen from the ranks of maintenance surveyors by virtue of their experience in the performance of buildings.

Education and training

A general view

(a) Education and training for building maintenance continues to follow a pattern which has now come to be considered as 'traditional'.

(b) Separate training of craft operatives for 'building maintenance' operations does not exist. Craft operatives follow courses of study at colleges of further education or similar educational establishments to obtain City and Guilds qualifications at Craft and Advanced Craft level in traditional building crafts.

(c) The Business and Technician Education Council's (BTEC) study programmes cater for technicians employed in the building maintenance sector.

Craft operative training

(a) Initial craft training for the traditional building trades is operated through the Training Agency's Youth Training Scheme. The Construction Industry Training Board (CITB) scheme currently requires the trainee to undertake a period of training of two years duration within which the trainee is required to carry out 'off the job' training at the local technical college/college of further education or training centre, and 'on the job' training with a 'sponsor' company.

(b) Attendance at college, which is usually by 'block release' pattern, will involve the trainee in completing a programme of standard practical exercises and development of trade skills which are devised by the Training Agency and appropriate to the specific trade being followed. All trainees are provided with a 'Training Record' in which a continuous statement of performance is maintained.

(c) At the completion of the initial two-year training period, the trainee will be expected to obtain an appropriate City and Guilds of London Institute Craft Certificate qualification.

(d) Apprenticeships are normally expected to be offered after an initial period spent by trainess with 'sponsor' companies, with whom they are placed by CITB, after they have completed their 'off the job' training period.

(e) After completion of the initial two-year period of training, the craft operative will be expected to continue studies to obtain the City and Guilds Advanced Craft Certificate and this will normally involve attendance by 'day-release' or block release at college, which normally takes one year of further study to complete.

(f) City and Guilds study programmes for craft operatives contain specific requirements to consider the implications for work of building maintenance. For the brickwork and masonry qualification there is reference to 'the causes and condition of work of alteration and repair'; for plastering 'repair and maintenance'; in plumbing, consideration has to be given to the 'technology of maintaining, repairing and maintaining installations' whilst for carpentry and joinery work there is reference to 'repairs and maintenance' in the craft certificate and a specific qualification at Advanced Craft level in 'maintenance work'.

(g) 'Skills Testing' procedures are now being operated at strategic locations to permit craft operatives to obtain 'qualified tradesman' status at about the time that they complete their Advanced Craft studies. This procedure is accepted by trades unions and employers as the point at which sufficient trade expertise has been demonstrated by the operative to justify payment of wage rates at the appropriate level.

(h) There exists a growing awareness of the need for continuing training at all levels of employment in building maintenance, including craft training and in the use of special tools and techniques. Although there is awareness of a need for multi-skill training, which has considerable advantages for building maintenance work, there is no formal training programme at present other than isolated courses at certain colleges offering specialist training in the use of special tools. Although many operatives acquire an ability to undertake multi-trade tasks, this is usually by taking one or two day courses arranged and paid for by the employer or by the operative attending evening classes.

(i) No 'specialised' training of craft operatives in techniques and processes involved in 'building maintenance' beyond these levels appears to currently exist. Although it is possible to proceed to higher levels of technical qualification, these are normally tailored to suit 'technicians' and the 'site supervisor' with little or no direct reference to the particular needs of the supervisor responsible for maintenance, repair and refurbishment of existing buildings. The Certificates or Diplomas in Site Management and the Diploma in Building Maintenance Management granted by the Chartered Institute of Building are such awards.

Technician training

(a) Education and training of technicians for the building industry is currently provided by BTEC with Diplomas for full-time study and Certificates for part-time study, at First, National and Higher National levels.

(b) Students are introduced to the general concept of the deterioration of buildings, and the need for effective maintenance, in 'units' of study concerned with building construction, science, materials, organisation and procedures.

(c) Particular coverage of building maintenance as a specific study can be contained within study programmes for the Higher National Certificate in Building Studies which can include units titled 'Assessment and Analysis of Buildings' and 'Maintenance and Adaptation of Buildings'. Some colleges have developed alternative building maintenance units for inclusion in such study programmes.

Alternative study patterns

(a) For smaller firms, which are essential to the well-being of the industry, management must be improved by continuing professional development on the part of individual members of the management team. This can be achieved by attendance at seminars and short study courses which are provided by educational establishments, professional bodies and commercial/industrial organisations.

Building maintenance management

(a) As building maintenance management is now regarded as a separate, important activity in its own right, this has led to the identification of specific needs and various courses are offered at educational establishments in a number of locations.

(b) After completion of a course of study for a BTEC Higher National Certificate, the student can proceed to a course of study for the Chartered Institute of Building's Diploma in Building Maintenance Management. Courses of study are provided at colleges throughout the country for candidates for the Diploma, although the Institute will normally require the candidate for examinations to have achieved at least seven years appropriate experience after the age of 18 or have passed either the Associate examination or an HNC in Building Studies and have three years acceptable experience in the specific sector of building maintenance.

(c) Subjects of the Diploma examinations are Maintenance Technology, Maintenance Management and Law and Regulations.

(d) Achievement of the Diploma in Building Maintenance Management will partially qualify the award holder for Associate membership of the Chartered Institute of Building, if not already in this (or a higher) grade of membership.

(e) It should also be noted that examinations for corporate membership of the Chartered Institute include papers in which the building maintenance specialist will find questions directly related to his specialism.

(f) There are no undergraduate programmes, other than a part of the BSc Estate Management Course, directly concerned with the building maintenance or maintenance management function, although study programmes will have reference to these specialism at some universities and polytechnics.

(g) At post graduate level, there is limited opportunity to obtain higher degrees which are specifically related to building maintenance.

Design/maintenance relationship

It is essential during design and construction that the client and the design and construction teams understand the requirements of the proposed building. At the design stage, decisions are taken which will affect the future use of the building and at this stage those who can make a contribution should be consulted.

The use of contractors who provide design and construct services, may prove to be economical to the building owner. The 'package deal' can be satisfactory when the job is straightforward and time is the main factor.

Action by the client and his staff

The client should:

(a) when selecting his architect/design team for the new building bear in mind the quality and performance of buildings they have previously designed;

(b) supply the architect with a full brief of his requirements for the new building, following consultation with his own maintenance manager;

(c) set up a small team from his staff to develop his ideas and requirements and to co-ordinate with the design team during the planning stages;

(d) allow the design team sufficient time to consider all his requirements and prepare well reasoned schemes;

(e) state his policy towards maintenance (capital/running cost relationship) and provide the architect with details of components standardisation, particularly electrical and mechanical equipment, which he requires;

(f) include in the brief his requirement for a maintenance manual.

Action by the architect/design team

The design team should:

(a) when studying the brief, take account of the client's maintenance policy and standardisation requirements;

(b) when no detailed brief is provided, obtain from the client's staff the necessary basic information to enable them to proceed with a composite solution;

(c) design the building to meet the client's functional needs, with due regard to the future cost in use;

(d) aim to consult the maintenance manager during the design stage on those areas where maintenance problems are likely to occur;

(e) before invitation to tender, ensure that the specification and drawings are vetted by the maintenance manager;

(f) be selective in its choice of materials for external use and their correct jointing;

(g) endeavour to see that when new, untried materials are to be used they will not fail in use and that any special manufacturer's instructions or recommendations on fixing etc are known to the contractors;

(h) supply as-built drawings and maintenance manuals to the client on completion or before if possible.

Action by the contractor

For certain types of building it is not uncommon for the main contractor to carry out only a very small proportion of the total work with directly employed labour, the remainder being carried out, under his responsibility, by specialists or sub-contractors. In order to reduce maintenance in this situation it is essential that the main contractor ensures that:

(a) all site staff, particularly site management, are well qualified and experienced in the type of construction being used;

(b) site management sets the quality of construction from the outset and ensures that this is achieved by all concerned;

(c) trades foremen check the specification and query any which appear to fall short of the accepted standard and reject inferior work by workmen under their control;

(d) site and supervisory staff are encouraged to question any item of design which they consider as a result of previous experience, could result in abnormally high maintenance or running costs.

Action by the maintenance manager

The maintenance manager can play an important part in influencing future maintenance costs of a new building by:

(a) helping in the preparation of the brief and by being involved with the team supervising the building programme;

(b) vetting the design as it is developed and before the scheme goes out to tender to ensure that the brief has been met;

(c) supplying the design team with information on the site conditions and sources of utilities;

(d) assisting in preparing a cost-in-use forecast;

(e) giving the design of standard fittings or other items, such as electrical and mechanical equipment;

(f) providing the design team with facilities (where applicable) to test materials in similar situations to those they will occupy in the new building;

(g) giving advice on what essential information is required in the maintenance manual.

Professional maintenance management

The increasing sophistication and complexity of buildings, their construction, finishes, fittings and services has led to the recognition that both day-to-day and long term operation of premises now requires dedicated, professional management. Not only are the buildings and their contents increasingly intricate but there are now considerable financial, legal and organisational constraints affecting every aspect of the site, structure and occupancy.

The growing capital cost of new buildings has also increased clients' awareness of the importance of having efficient and cost effective designs and well managed building contracts. Nevertheless existing buildings tend to be neglected despite their importance as part of a national capital asset. This presents a great challenge to those responsible for maintenance and calls for a professional specialist maintenance manager.

The task of maintaining the existing building stock is wide in scope and has to be carried out against a background of restraints. Often, buildings cannot be closed for operations even for major works, and this calls for a high order of skill and judgement from the maintenance manager to achieve his goal without disrupting others.

The maintenance manager has a commitment throughout the life of the building. It is paradoxical that professional status and recognition are automatically accorded to the designer, whose involvement with the building is relatively short, invariably ending with the completion of the building, and yet often denied to the maintenance manager whose commitment lasts for the lifetime of the building. The commitment implies a full understanding of the building and its characteristics. Not only must the manager maintain it, but also alter, extend and convert it.

As more clients come to count the escalating costs of maintaining their buildings, it is inevitable that designers will increasingly seek advice from maintenance managers to avoid those features of design which add to the client's maintenance bill.

To be effective, a maintenance manager must have a broad outlook and specialist knowledge equivalent to degree standard and this knowledge must be rounded off by practical experience to give him expertise in the following:

(a) property registers–surveys and record systems;

(b) maintenance systems for buildings and services–day-to-day, emergency and planned maintenance;

(c) diagnosis and problem solving–materials, technical and economic, including cost in use techniques, appraisal and failure patterns;

(d) administration routines covering the office, the single project and multi-project programmes;

(e) planning and control systems covering physical and financial aspects such as directly employed labour (DEL) including bonus schemes and forecasting maintenance budgets;

(f) alternative methods of contract execution;

(g) legislation relating to contracts, property, working conditions and employment;

(h) standards–maintenance, service times, design, brief writing and specifications;

(i) maintenance feedback schemes–building and engineering systems;

(k) audit functions–quality, response times, economy, procedural and statutory compliance.

MAINTENANCE MANAGEMENT

These technical and professional skills must be balanced by a commensurate ability in general management and practical expertise calling for the development of specific dynamic skills:

(a) interpersonal skills–leadership, delegation and motivation styles;

(b) communication skills–the effective use of written, oral and visual media, including chairmanship and public speaking;

(c) the art of science of risk analysis and decision making, including the setting of personal and organisational goals;

(d) the technical criteria for the effective utilisation of people by an organisation–job analysis, specification, recruitment and training policies, including industrial relations;

(e) the definition, creation and maintenance of appropriate management information systems.

The professional maintenance manager must also be prepared to deal with such matters as:

(a) land utilisation, including landscaping, access and boundaries;

(b) leases, including tenure, covenants, rents, repairing obligations and dilapidations;

(c) town planning matters, including development control, appeals and enforcement;

(d) rating, including assessment appeals, allowances and exemptions;

(e) all services and utilities, including their statutory requirements;

(f) all environmental matters affecting the occupancy;

(g) cleaning and day-to-day operation of the premises;

(h) health and safety at work matters;

(i) fire safety and practical as well as statutory requirements for fire prevention and means of escape in case of fire;

(j) security against intruders or vandalism with particular regard to the protection of the property, contents, occupants and information;

(k) insurance, including buildings, contents and liability;

(l) budgeting including inspection, planning, estimating, costing, monitoring and controlling.

In addition to the above his services will probably have to include energy conservation matters, refurbishment and upgrading works and relocation requirements.

In short, the time when the maintenance manager's main duty was to deal only with interior and exterior repairs and decorations is long past, although the maintenance of the structure itself must still form the most important task within his remit.

The professional maintenance manager must be sufficiently involved with the affairs of his employing organisation to acknowledge that his maintenance policies must be consistent with the main objectives and financial state of that organisation and that there is no conflict which may adversely affect production, staff or finances. The owner or owners of the organisation are the ultimate arbiters of each organisation's way of rapid and effective diagnosis and of obtaining a correct specification and a high level of workmanship.

The effects of the standards of maintenance employed will then be studied in the light of the operational efficiency of the organisation and the morale of the staff.

Motivation and measure of performance

Manpower is expensive and must not be wasted. The workload at any given time should be measured to calculate manpower requirements. In new construction the work analysis or work plan gives guidance on the manpower and trades required. In repairs and maintenance the workload is measured by trade and the manpower budget calculated by considering the forecast number of repair jobs per year, the work content per job in productive standard hours, the effective performance of operatives, the percentage time lost for sickness and leave and then converting to man years. Maintenance and servicing work is labour intensive, the work fragmented and scattered and often very difficult to supervise and control.

Production performances are difficult to measure and achieve unless some form of work measurement, coupled with output targets, is used. The workmen, being the medium through which work is performed, are the most important factor in performace. Adequate financial rewards must be provided for good performance and for good quality control. The setting of realistic time targets, effective work planning and promotion of pride in work should have high priority.

Productivity can often be doubled by adopting these methods, resulting in good take-home pay, with considerably reduced unit costs. Money is a great motivator but there are other factors which help performance –good labour relations–good working conditions–communications and relationships with others–sense of achievement–good supervision and enlightened management.

The discipline of MBO is a good motivator for supervisors and managers. It is achieved by setting key target areas for planning, progress, improving performance, cost effectiveness, financial management and budgetary control.

Motivation must be seen to initiate action at all times, to prevent a diminishing effect on incentive schemes. Bonus drift must be prevented and one must aim constantly to improve methods (there is always a better way of doing things), update time standards, reduce non-productive time, reduce unit costs and increase earnings for increased effort.

Bonus drift can be prevented by the regular monitoring of bonus schemes. This must be achieved by continuous weekly control information relating bonus to productivity and to the average work content per job. There must also be physical work checks undertaken both by supervision and by officers not directly controlling the labour force but with a thorough understanding of the bonus schemes. The values must be subject to continous review with a full comprehensive revision every 3-5 years. Work specifications setting out the conditions under which the scheme operates must include provision for these regular reviews and revisions.

Drift occurs when there are:

(a) unauthorised changes in working methods or practices;

(b) reductions in the standards of quality or service below those used as a basis for the time values;

(c) overclaims in the quantity of work done;

(d) ill defined work values and as a result, employees can claim additional time values which overlap the basic job value;

(e) overclaims for non-productive allowance;

(f) inadequate job descriptions giving scope for overclaiming of work;

(g) needs to test the technical assessment of work of operatives (eg. fault finding in electrical defects).

The building user must be encouraged to describe maintenance correctly and the maintenance manager must use that information to identify the correct operative or contractor. The operative or contractor must then arrange access together with the provision of correct tools and materials to carry out the work swiftly and effectively, thus avoiding any repeat visits.

Feedback

The rectification of problems found to result from incorrect detailing, poor specifications and choice of materials, defective construction techniques and workmanship is very costly. Therefore, it is prudent for these defects to be recorded and the necessary remedial measures adequately described.

This information can be of great value when fed back to the original designers if they ensure that the message is received, understood and acted upon.

Maintenance management will be aware of good design features and of materials and fittings which stand up well to everyday use and require little or no maintenance. These matters should also be recorded and fed back to the designers.

Standards for building materials and components issued by the British Standards Institution (BSI), are set at a minimum level. If, after an appropriate time in use, any standard is found to be inadequate then BSI should be informed.

Information and data collected by maintenance management which could be of value to others, should be disseminated as widely as possible by lectures and publications.

THE ROLE OF THE COMPUTER

The advance of the technology

Until the early 1980s, computers in maintenance were confined mainly to a handful of organisations requiring a standard system for general application on a large scale. These were usually systems developed from scratch for whichever department had the responsibility for carrying out day-to-day or jobbing repairs, using a centrally based computing facility that served the whole organisation.

In the public sector, systems for process jobbing repairs and works orders became more widespread, particularly where local authorities had a reasonably sized housing stock and employed a direct labour organisation (DLO) to carry out the work. Cheaper computing costs and improved communications capabilities enabled some of the more monolithic systems to be decentralised so that they could be operated from intelligent terminals in locations away from head office.

In the last few years the pace of change in computing technology has accelerated to such an extent that it is safe to forecast a dramatic increase in the use of computers in the field of maintenance management in all its aspects, in the private sector as well as the public, and for the small as well as the large property owner.

It is the recent developments in software that offer the greatest promise from the prospective user's point of view. Not only are there more maintenance packages available generally, but the new general purpose software enables users either to generate their own applications, or to have them written quickly and inexpensively to their own specification by experts in the use of the systems.

The most powerful of these general purpose packages are known as 'relational databases'. A 'query language' enables the database to be held and retrieved from a series of open files, simplifying file handling routines and increasing the flexibility of the database and the user's ability to generate ad hoc reports.

Maintenance is ideally suited to this approach, because it is essentially about the recording and retrieval of vast amounts of data, from the initial components from which a building is constructed, to its place in the property portfolio, its tenants, its repair history and all the different aspects of its service life. The sheer size of the property maintenance information base and its lack of clear definitions and classification has been one of the main reasons for the lack of computer applications in this area.

New technology is not limited to information retrieval and the generation of reports. Micro-computers may be linked to mainframes, providing input to existing applications such as repairs reporting, or operated as stand-alone systems doing everything from recording condition surveys to running cyclical maintenance programmes. Modules in the general purpose software packages can be used for word-processing or as spreadsheets to prepare budgets or run financial models for cost-in-use analysis.

The attractions of this situation for the maintenance manager are many and real and for the smaller organisation in particular, there now appears to be a clear opportunity to justify computerisation, on the strength of the range of applications that can be covered.

Future uses of computers in maintenance management

Most of the readily identifiable areas for computer application are common to both large and small property portfolios. They include:

(a) maintaining a property database;

(b) procedures for dealing with day-to-day repairs;

(c) systems for cyclical and planned maintenance;

(d) budgetary control and financial modelling.

Property database

The main purpose of a property database at the upper end of the scale, eg a large housing authority, is to keep records of the condition of the stock, in order to maximise the effectiveness of its maintenance effort through programmed repairs, rather than piecemeal repair and replacement. The property file should contain, for each dwelling, constructional details, the remaining life of the principal components derived from condition surveys and an estimate of their replacement costs.

A further use for the property database is to provide the main file in the repairs reporting system, where it will be integrated with information about tenants and other details drawn from lettings and rents systems. Most property files in housing authorities at present take this form and are used additionally to record the 'repairs history' of each dwelling, so that details can be checked of both current and past repair work on a dwelling or groups of dwellings.

For the property owner with a mixed portfolio, the property database is more likely to begin life as an asset register, but there is equal justification in extending its content to include details of construction elements within individual properties and to use the file as an aid in assessing replacement need and preparing longer-term maintenance budgets.

The principle of the property database can be extended to facilities management functions where large individual properties or groups of properties on the same site may be concerned. Whether they are commercial properties, hospitals, universities or museums, they will have the same essential constructional features of fabric, services and finishings to be maintained, as well as additional items such as energy monitoring to be considered. Here there is a case for the property database to include more detailed information about room layouts, finishes and mechanical and electrical installations and their performance, to improve information retrieval facilities and reducing reliance on other forms of documentation.

Maintenance managers have generally been very poorly served in the past by the designers of new buildings, in terms of as-built drawings and data on the expected performance of the various installations in use. There are now signs that building owners are becoming much more concerned to know how their buildings are going to perform and this will mean that designers will not only need to take this into account when cost-planning, but will also be obliged to produce more accurate records of the completed building and its services, as well as comprehensive maintenance manuals.

Day-to-day repairs

Day-to-day repairs represent the biggest single field for the application of computers in building maintenance, because of the many benefits to be gained, from mechanising what is a time-consuming and expensive administrative exercise. Systems are becoming more attuned to the various needs of client departments and are also more consumer orientated. Repairs reporting systems now have improved enquiry facilities, to tell tenants the status of the repair they have ordered, and many authorities now issue tenants with repair receipts and customer satisfaction slips for return in the event of the work being unacceptable.

The repairs system is also now used by most local authorities not only to log repair requests, but also to administer term contracts for jobbing repairs and repairs to void properties, whether the work is carried out by their own

DLO, private contractors or a combination of the two. The more advanced systems incorporate schedules of rates, to enable works orders to be printed bearing appropriate descriptions of the required repairs and for use in budgetary control and processing invoices. Where schedules of rates are used, works orders can be priced automatically for commitment accounting purposes and details held on file for use in checking invoices rendered by contractors. If a DLO is used, the DLO Revenue Account can be credited automatically with the value of the completed orders.

A logical extension to a jobbing repairs system, particularly where there is a direct labour organisation (DLO) is for non-urgent repairs to be scheduled so that they can be carried out in a planned way, maximising the efficiency of the works organisation. Such an approach is a feature of the cyclical or zoned repairs systems operated by a number of the larger public authorities. The property stock is divided into zones and a regular cycle arranged for repair gangs to carry out works orders accumulated between visits. The computer can be used to program the work for each gang, prepare lists for pre-inspection, and, from the inspections, prepare lists of materials for arranging deliveries when the gang is on site.

Many DLOs enjoy direct links to their client department's repairs system, making it possible to computerise many aspects of their operations, including production control, wages and bonus payments, stock control and financial accounting. There is no reason why private contractors specialising in maintenance work should not adopt similar systems. They would not only benefit from improvements in their own efficiency, but would also be able to offer a service that could link up with the client's own computer system. For their part, a number of local authorities are already examining ways in which they can offer facilities to private contractors to communicate with their systems direct, such as the printing of works orders at the contractor's office and the input of completed works order details prior to the submission of invoices.

Cyclical and planned maintenance

Cyclical maintenance consists primarily of setting up and administering programmes for the decoration of buildings, as well as contracts for the regular servicing of mechanical and electrical installations, including lifts, fire appliances, alarms and the like. Computers have had little impact in this area to date, except in the fairly specialised field of mechanical and electrical plant maintenance, where simple lists of equipment can be drawn up and schedules of required work programmed and brought forward as necessary.

One of the potential benefits of a comprehensive property database is the facility to record the dates when elements were last inspected and the date when orders need to be placed or renewed for the next round of servicing or redecoration. The file can be used as a diary to extract lists of items requiring attention in date order. A record can also be kept for cost analysis purposes, of all orders placed and actual expenditure against individual items.

There is considerable scope for extending the use of computers in relation to painting contracts, because of the repetitive nature of the work, where the same numbers of dwellings and quantity of work come round at each cycle. A computer-based system could keep track of dwellings sold since the last cycle, or where the wooden windows had been replaced with uPVC, but again the system needs a property database in order to gain the full benefits of such a system.

The same is true for major repairs and replacement (planned maintenance) contracts, which at present are largely drawn up as the result of ad hoc inspections or the identification of problems on a specific building or estate. These are brought into the capital expenditure programme as necessary, carried out when funding permits, but at no time are records of the work planned or completed stored on a property file for future retrieval. Much of this will change as property owners adopt comprehensive databases that enable their maintenance managers to plan ahead and keep full and readily accessible records of what they do.

Budgetary control and financial modelling

In the next few years, one of the areas of most rapid growth in the maintenance field, as in many others, seems likely to be in the use of spreadsheets. They can be used to carry out a wide range of tasks, from the preparation of simple budgets to complex financial modelling. Multiple spreadsheets can be used to aggregate data, for example in creating a corporate budget from a set of regional figures.

Already practical examples can be found of some of the more obvious applications, in budgetary control and the preparation of profiles of future maintenance expenditure on parts of a housing stock or specific elements within it. At the experimental level, spreadsheets are being used to construct models for the following purposes:

- evaluating costs during the early design stages;
- evaluating property maintenance strategy;
- life-cycle costing;
- investment appraisal.

MAINTENANCE MANUALS

The importance of handing over information, drawings, schedules, etc at the completion of a new project (be it a new building or an alteration or addition) to the building owner, user and maintainer of the building cannot be over emphasised. This information is best produced in the form of a loose leaf manual, sometimes referred to as the project information manual. The manual should contain all essential information about the construction of a building, together with all the necessary sources of information for its proper maintenance. Only special or unusual maintenance treatments should be included. Particular attention should be paid to information regarding safety and fire integrity, structural design loading and fire escape routes. On an existing building, where no manual has been provided, it is recommended that the maintenance manager should take steps to produce one, using as much information as can be mustered.

The manual can:

- foster a better architect/maintainer relationship;
- enable a property to be more effectively maintained;
- encourage planned maintenance;
- make it possible to avoid misusing a building or elements of a building;
- ensure the continuity of the design intentions or capability of the building and its services, throughout its life.

On large projects it may be necessary to produce a separate manual for each department or phase of the scheme. All essential 'as-built' drawings should be provided with the manual. A drawing showing fire compartmentations, escape routes, etc, should also be included. Such a manual could be compiled in the following sections:

(a) Contract use and legal aspects.

    (i)   general description of project: construction; number of storeys; floor areas; ceiling heights, etc;

    (ii)   contract information: parties to the contract and type; dates for possession of site, practical completion and defects liability period; schedule of defects; certificates of making good defects;

    (iii)   contract consultants: names, addresses and telephone numbers of design team and their consultants;

    (iv)   public utilities: names and addresses of contacts at local authority and all public utilities;

    (v)   consents and approvals: statutory requirements for which certificates have been obtained;

    (vi)   easements and covenants: rights of way or any other restrictive covenants;

    (vii)   design loading: essential information about structural elements;

    (b) Building elements, finishes, fittings etc.

    (c) Site information.

    (d) List of 'as-built' drawings provided.

    (e) Maintenance contracts.

    (f) Guarantees.

    (g) Special maintenance items.

    (h) Additional notes.

It should be borne in mind that the object of a manual is to provide useful information to the client organisation and regard should be paid to the expertise which already exists, ie. it is not necessary to provide information, for example, on how to maintain standard doors, windows, etc.

Chapter 6
Financing and Budgetary Control

COMPONENTS OF MAINTENANCE EXPENDITURE

To a building owner the maintenance of premises usually means the expenditure incurred annually for the premises. For the purpose of this guide, expenditure is broken down into four main groups:

Group 1: rent, rates, taxes, insurances, etc

(a) rent or the capital amortised over a term of years;

(b) rates;

(c) taxes;

(d) insurance.

Group 2: utilities

(a) gas;

(b) electricity;

(c) water;

(d) heating fuel (coal, gas, oil, electricity);

(e) compressed air and other specialist services.

Group 3: repairs to the fabric and engineering services and grounds

(a) building fabric, including decoration, drainage, etc and services;

(b) furniture and fixtures;;

(c) electrical services including lifts, lamp changing and cleaning;

(d) technical, heating and ventilation services, including those demanded by the Factories Acts and other legislation;

(e) tools, plant and scaffolding;

(f) renewals of large equipment, rewiring, reheating, etc (to this item is sometimes added alterations or additions, improvements and minor works);

(g) grounds and environment, including weed control and horticulture.

Group 4: ancillary and housekeeping services

(a) porters, messengers, etc;

(b) security, day or night, including car park attendants;

(c) cleaners and cleaning materials and repairs to equipment;

(d) window cleaning and laundry;

(e) fire precautions and safety;

(f) waste disposal;

(g) telephone and fax.

(Sometimes post and stationery are included)

Whilst it is essential for most maintenance organisations to budget annually, because accounts are by law certified on this basis, it is considered more efficient to operate on a longer financial management period of three to five years. Some organisations operate an accounting system which closes the expenditure as at the last working date of the financial year, ie. all wages and invoices paid go into the accounting year. Other organisations operate so that all invoices processed over, say, two weeks after the final date, are charged back. In addition, maintenance managers make a valuation of large jobs, deduct payments made and give the accountant the value of work done, but not paid for, as a debit. There are arguments for both methods but the second provides a more accurate record. Commitment accounting is often used in the financial management of maintenance.

There are organisations in which additional finance is found from some sub-heads of expenditure and offered to the maintenance manager a few weeks before the end of the year. This is *not* a sensible way of utilising available funds. Long term planning and more flexible accounting systems are a better way of dealing with this. However, if 'gifts' of money are inevitable, then how such windfalls should be spent must be pre-planned. Even when operating within a long term forecast, say five years, it becomes necessary to update forecasts annually. Today, costs change so rapidly that it is necessary to examine the actual expenditure on a monthly basis, or more frequently.

LONG TERM FORECASTING FOR REPAIRS AND MAINTENANCE

Many organisations calculate the annual maintenance budget on the basis of the previous year's expenditure, plus a factor for inflation and growth. This is unsatisfactory and economically unsound, where the intention is to maintain adequately the organisation's major capital asset. In particular, this method does not take account of the evaluated workload and servicing arrangements.

The evaluated workload is the total work for any one year which is calculated to meet the maintenance service standards or policy laid down by the organisation and to prevent the properties from falling into a state of disrepair.

As an aid to effective maintenance, it is important to establish an annual budget which reflects workload requirements and expresses these on the basis of an acceptable level of labour performance and 'value for money'.

The use of production engineering, cost control, management techniques, such as MBO and budgetary control procedures, are invaluable in achieving effective financial management and the planning of the workload requirements, which in turn are essential pre-requisites in achieving the goals laid down for the maintenance plan.

AN EXAMPLE OF THE PREPARATION OF ANNUAL ESTIMATES

The best way of illustrating the principles used to prepare annual estimates is to take the estimates for a group of buildings as in the following example:

Annual estimates are considered for part of an estate of mixed buildings with an average age of 30 years and 300,000^2 gross floor area. The replacement value of the superstructure, with no site cost, say at £150 per m^2, is £45m.

The estimate is prepared under four main sub-heads of annual expenditure. (This is an example only and cannot be financially related to current buildings):

(a) building, fixtures and furniture;

(b) electrical;

(c) mechanical, tools and plant;

(d) renewals.

Building, fixtures and furniture

		£
(i)	redecorations on programme	270,000
(ii)	service contract for door springs.	1,500
(iii)	specialist contractors, maintenance work such as metal windows, blinds, stone and terrazzo, locks and keys, floor finishing. (In this group are separately identified jobs each estimated to cost less the £250)	37,500
(iv)	major repairs to roofs, roads and car parks, floors, walls, etc (each costing over £250), derived from the area surveyor's annual inspection list and in accordance with priorities.	90,000
(v)	day-to-day repairs of a minor nature (majority £50 or less).	129,000
		528,000

Electrical

(i)	lifts service contracts (44 lifts).	33,000
(ii)	lift repairs.	24,000
(iii)	service contracts, batteries for fire alarms, emergency lighting, clock systems.	3,000
(iv)	small contracts (each costing less than £500).	7,500

Total c/f 595,500

			b/f
			595,500
(v)	major contracts, sub-stations, switchgear etc, rewiring etc (between £750 and £1,500).	21,000	
(vi)	lamp changing.	22,500	
(vii)	lamp fitting, washing and cleaning (once per year).	24,000	
(viii)	day-to-day repairs.	135,000	
			202,500

Mechanical–including tools and plant

(i)	service contracts for boiler burners, electronic boiler controls, refrigeration plants insurance, inspections of hoists, lifting tackle and pressure vessels.	25,000	
(ii)	specialist contracts (less than £500) such as ductwork, filters, hose-reels, lagging.	24,000	
(iii)	major repairs (over £500 and less than £15,000) prepared from inspections and records of boilers etc.	34,500	
(iv)	tools and plant.	12,000	
(v)	day-to-day repairs.	175,500	
			271,000
	Total of building electrical and mechanical		
			1,069,000

Renewals

(i)	building–roofs and floors.	45,000	
(ii)	electrical – one lift and other installations.	135,000	
(iii)	mechanical–boilers (small) and other installations.	75,000	
			255,000
			1,324,000

In addition, the changes in labour costs, both on the hourly rate and the overheads, materials, plant etc, should be worked out accurately at intervals. By taking both into consideration, the maintenance manager can obtain an accurate forecast.

DEALING WITH OVERHEADS

An efficient maintenance manager will examine the cost of running his department, including any departmental building(s) and its (their) upkeep at least once a year. The maintenance manager should examine his costs in the following broad categories:

(a) stores and materials

(b) labour;

(c) professional and technical services.

Stores and materials

Where the owner's policy is to hold some materials in store there will be a capital outlay on facilities and on the materials themselves. In cases where the supply of materials is subject to delay then a stores holding is essential.

There is a great need to plan the maintenance workload so that suppliers of essential materials can keep pace. Where high interest rates prevail it is not economically viable to hold materials in store for long periods. If the building owner requires a quick response by the maintenance department, because loss of a machine or other facility would cause a loss of production and profit, then it becomes necessary to hold stocks of materials most likely to be required. The maintenance department stores may be used as the main stockholder of such items for other departments of the parent organisation. The size of the stores and its staffing will depend on the amount of stock held.

DIRECTLY EMPLOYED LABOUR–PROCEDURES FOR ESTABLISHING THE MANPOWER BUDGET

Work study engineers forecast the direct labour manpower budgets by trade in advance of the preparation of annual estimates. The manpower budgets are then reviewed six months later to allow adjustment in the light of the latest work trends, prior to the coming financial year. The procedures which will now be outlined can be used to establish the direct labour manpower budget as part of the financial estimates.

Basis of calculation–all trades (excluding painters)

Incoming workload–repairs and maintenance.

Those responsible for work study should maintain records of incoming numbers of repairs requests and estimated workloads. These records, which may cover a number of years, give a reliable indication of the trends past and present and thereby a rational forecast for the future. Greater weighting is given to more recent results of performances and any major trends are carefully investigated. The budgeted number of jobs forecast for a coming year must take into account the following factors:

(a) Changes in numbers/types of units comprising the estate.

Generally, the number of jobs is directly proportional to the increase or reduction in the number of units to be maintained, although the types of properties and services must be taken into account, eg. the purchase and construction of new buildings and the acquisition of new equipment.

(b) Maintenance service standards.

Maintenance service standards should be agreed for each financial year. These standards specify the servicing requirements and average response times in each trade and the maximum acceptable job completion periods. (Examples are given at Appendix C). Changes to these standards may affect the likely number of jobs to be completed each week. Legislation, such as the Health and Safety at Work etc Act, the Defective Premises Act, Housing and Factories' Acts will also have some influence on standards and will need to be taken into consideration.

The maintenance organisation will need to set out a number of routine maintenance programmes which must be completed each year, records being kept of the numbers of units covered by each programme. These should be checked by those responsible for work study to establish the number of jobs for routine maintenance work.

(c) Variations in the incoming number of jobs.

There are variations, week to week, in the number of incoming jobs requested by occupants and from other sources due to seasonal factors and emergency situations. In practice, these fluctuations, when low, can be offset by feeding to the work force programmed routine maintenance and preventive maintenance tasks. Peaks in the workload which cannot be programmed economically for the labour force can be covered by using contractors. Annual estimates should include a provision for this.

To obtain the basic annual cost of each class of operative, the weekly cost is multiplied by the estimated number of weeks of productive work, ie. excluding annual and sick leave and public holidays in the year.

Overhead charges

Overheads are added to the labour cost as a percentage addition calculated to cover the following items:

(a) holidays and sick pay;

(b) National Insurance, superannuation and death grants;

(c) rents and rates of works accommodation;

(d) purchase, lease or hire of vehicles, plant and mechanical tools (including maintenance and repair);

(e) establishment expenses and charges by other departments;

(f) gratuities (eg. ex gratia payments)

(g) return on capital employed.

These components of the percentage addition are monitored monthly from costing information and adjusted to ensure an adequate recovery rate.

Establishment and staff on-costs

In addition to costs included in the overhead charges, a further percentage addition is made to cover the salaries of administrative, professional, technical, executive and clerical staff employed on maintenance management activities.

Materials

The estimated cost of materials is calculated in conjunction with materials management operations on the basis of current expenditure. The total cost for budgeting is spread over the various trades in proportion to the estimated volume of materials useage per trade or activity.

Total cost of the labour force

The basic cost per year of each trade, with the addition of overheads, on-costs, and materials costs, is multiplied by the number of staff in the manpower budget to produce the total estimated cost of the labour force. The total cost is allocated on the basis of the estimated workloads, to all activities of estate maintenance, including special maintenance, upkeep of gardens and sports grounds, ancillary improvement work, repairs and improvements to all properties, car parks, roads, drains, sewerage works etc.

ESTIMATED COST OF CONTRACT WORK

Due to fluctuations in the total workload any direct labour force is unable to deal economically with more than about 75% of the overall workload. To accommodate the deficiency, work is put out to contractors. Contract work may be placed into one of three main categories:

(a) trade support to cover peak workloads or shortages of operatives;

(b) local contracts for specialist work, eg. asphalt roofs, roads etc;

(c) major contracts for large works for which it would be unreasonable to use directly employed labour.

(a) and (b) are usually covered by term contracts, whereas major contractors (c) generally cover all major maintenance jobs, remedial and improvement work. Estimated costs for these are prepared for budgeting. A thorough investigation of all such work is necessary before drawings, specifications and estimates are prepared. These methods may also be used if the organisation has no directly employed labour force.

PROFESSIONAL AND TECHNICAL SERVICES

The appropriate part of professional staff salaries will be charged against direct labour.

The professional members of staff involved provide a service to the parent organisation of advice, estimating, preparing drawings and specifications, obtaining tenders, supervising the work in progress, checking contractors' accounts and certifying for payment.

The costs must, therefore, be calculated and apportioned between contract, direct labour and establishment charges.

ALLOCATIONS FOR THE FINANCIAL YEAR

The financial resources to be made available for the ensuing year are calculated and allocations made for all activities to ensure that maintenance service standards are met.

To comply with the resources finally made available, it may be necessary to amend these standards to ensure cover for the following:

(a) cost of the required labour force;

(b) cost of the contract work which has been agreed as necessary to cover the priority jobs in the three categories specified earlier. Work programmes are then prepared and agreed with senior management.

Even when the budget has been carefully allocated, care is necessary to ensure that the resources are used wisely and that the various allocations of finance are not overspent; some flexibility is acceptable to cover unforseen emergency work. A monthly programme of work should be prepared, including all priority work, together with financial estimates. These are used as the work plan and as a basis for budgetary control.

It is normal practice for the maintenance manager's forecast estimate to be sent for consideration by the appropriate authority, which often finds it necessary to reduce the estimated figure. Ideally, any reduction should only be made following discussions with the maintenance manager.

Assuming that a forecast of £1.5m is reduced by 5% to £1.425m, the maintenance manager then has to keep within this amount and must not overspend without written authority from the management board or council. The reduction in schemes or service provided to meet the approved budget needs to be clearly stated to the client, so that he knows what he is losing by his inability or unwillingness to provide the funds.

The maintenance manager and building owner will wish to examine the costs of running the maintenance organisation and compare them with the cost of independent consultants providing similar services.

One way of checking is to calculate, on the basis of fee scales in operation, the services of a partnership in an architectural practice, a building surveying practice and an estate agency. When calculating the fees due under this method it must be remembered that the work is in small units and also covers the engineering services; it should be weighted accordingly.

The standard of service to be provided will include preparation of drawings, specifications, obtaining tenders, supervision of work and certification of accounts. The fees on this basis are likely to be between 15% and 25% of the value of the work completed.

The maintenance manager must now look at this priority list for small contract works and major works for building and electrical and mechanical works; he can also look at the list of renewals. On examination he may be able to defer by approval sufficient jobs to meet the reduced budget. Tender figures below or above that anticipated are also taken into account.

COSTING AND BUDGETARY CONTROL

In section 7 the policy of long term forecasting is advocated. With an estate comprising buildings of widely differing ages, many of the major items of expenditure fluctuate year by year. Consequently, a good manager uses the long term forecast of work to even out the annual workload and the annual estimates to avoid violent fluctuations, which the building owner would find difficult to deal with and pay for. These major items of expenditure may be:

(a) redecorations (perhaps five years external and seven internal);

(b) repairs to roofs, roads and hardstandings;

(c) pointing to brickwork and stonework;

(d) renewal of lifts, electrical rewiring, upgrading of lighting to higher standards;

(e) replacements of mechanical equipment, upgrading heating or insulation.

FINANCIAL CONTROL OF CONTRACTS

Contract work must be efficiently supervised to ensure that the building owner obtains an appropriate return on his expenditure. Omissions and additions to contract documents must be fully documented and priced. Any variation must be covered by the issue of a variation order at the time it is found necessary–the rates in the contract provide the basis for adjustment. Where items are not covered by the contract rates, a quotation must be obtained from the contractor.

EXCESS COSTS

If the sum of the variations is likely to cause the contract sum to be exceeded, an immediate report should be made to senior management giving reasons and a detailed estimate. The contractor must not be instructed to proceed with such additional work until the necessary authority to continue has been received.

FINAL ACCOUNTS

Final accounts should be set out clearly indicating:

(a) contract sum;

(b) omissions;

(c) additions;

(d) sub-contractors' accounts;

(e) supporting receipts, accounts, VAT account etc;

(f) final sum.

It has been previously shown how it is possible to build up methodically a forecasting system based upon sound records and experience. This system is more applicable to larger organisations but the principles can be applied to any maintenance expenditure. It is again important to emphasise that it is not sufficient merely to take last year's figures and to add a percentage for any increases in cost of labour and materials.

Once the annual budget has been agreed, steps should be taken to put the programme into operation. It is good practice to have the budget approved in time for the maintenance manager to have specifications prepared and tenders accepted, to enable orders to be issued well before the commencement of the financial year. Service and term contracts need to be negotiated in good time.

Most large organisations have some form of costing system for the products or services they provide and the costing system for repairs and improvements will be often dovetailed into the main system. A costing system should not be introduced without an evaluation of its worth to the building owner. If a system is introduced then it should be simple to understand and operate as well as being capable of providing the basic information for management control. Legislation lays down the accounting procedure to be adopted in local authority works organisations.

Looking ahead over five years gives the maintenance manager an opportunity to assess commitments and thereby even out the expenditure according to the availability of funds. A prudent manager will always have a list of jobs which need doing when funds become available at short notice, or which rely on access being permitted to premises.

Large estate owners generally need to know the cost of repairs to individual buildings. The costing system should provide this information. Many maintenance managers act as agents, carrying out work for different sections of the owner's department and it then becomes necessary to show how much individual jobs have cost.

For most jobs it is necessary for the workman to write down the time he spends on the work, together with the details of the materials used. These two elements can form the basis of any costing system. If the man is given a works order for each job, which describes the work and indicates the code to which the work is to be charged, this can be used as a basis to obtain the data upon which to operate the system.

The maintenance manager needs some form of regular management control information so that he can see quickly whether his programme is running smoothly and according to the overall plan and also to assist him in the decision-making process.

Chapter 7

The Execution of Maintenance

WORKS CARRIED OUT BY CONTRACTORS

Lump sum tenders

It is necessary to furnish to the contractor an accurate description of what is expected from the tender. Depending on the scale and complexity of the work this may be formulated in a number of ways. For the larger works the information should be prepared by qualified staff, whether employed in-house or engaged as consultants.

Taking the replacement of windows to a large office block as an example:

The first requirement would be an initial survey with recommendations to the building owner regarding the most appropriate design and materials to be used, together with costs. With uPVC, hardwood, softwood, aluminium and steel windows available it is important that the advantages and disadvantages of the various types are set out for the building owner to make his choice, bearing in mind initial cost of replacement and subsequent maintenance, decoration, cleaning and repair.

When the building owner has considered the recommendations and made his decision he will instruct his surveyor to prepare tender documents.

For a project of this nature these will consist of a specification setting out the requirements in respect of workmanship and materials with details of the site, location of the site compound, any restrictions on working times, the necessity to comply with Acts of Parliament governing such works, with particular reference to the Health and Safety at Work etc. Act 1974 and any other relevant information or instruction. Drawings indicating the design of windows should be provided unless requirements can be defined accurately by reference to manufacturer's catalogue. The form of contract to be used should be stated and a form of tender provided.

The invitation to tender should be sent to at least three but no more than six contractors with a date for submission of tenders stated. The surveyor will advise on suitable firms to be invited.

Quotations

Some works, such as repointing or re-rendering of a boundary wall to a factory, may be satisfactorily dealt with on a quotation basis, with the contractor providing the specification. This requires an accurate description of the extent of the work to be carried out but allows the contractor to give a lump sum price on his own conditions. Since these may vary from one contractor to another it is important to consider each set of conditions carefully before deciding which to accept.

As a guide three quotations should be sought.

Term contracts on schedule of rates basis

This type of contract is used most where there are recurring items of work to be completed over an agreed period of time.

The basic document consists of a schedule of all the individual items or operations which are expected to be required during the period. Each item or operation will have a measurement unit against it such as cubic metre, square metre or quantity.

Schedules may be sent to contractors to price and return but since tenders may be difficult to compare on this basis it is more usual to have the schedule pre-priced by an independent estimator or quantity surveyor. In the latter case contractors will be asked to tender by quoting a percentage on or off the rates which the estimator has provided. In order that readily comparable tenders may be obtained a single percentage is only usually applied to each trade overall, or quite often only to the eliminator's total price of the tender. An assessment is made of all the unit items and each final trade percentage or the total percentage is divided by adopting a 'swings and roundabouts' approach to arrive at the mean percentage.

The term of the contract may be for one, two or three years.

Service contracts

Service contracts are used mostly for the repeat servicing of such items as doors, locks, shutters, air conditioning plant, alarms etc. The cost is predetermined for one year in advance after assessing frequency of visits, content of servicing and parts to be provided. Generally, extra call outs are catered for in the contract, except for damage attributable to outside influence.

Contracts should be obtained by competition and reviewed at least once a year to update schedules and to revise specifications.

Negotiated contracts

Lump sum contracts, term contracts and service contracts may be negotiated.

When a contract is negotiated, the contractor is often selected on the basis of past performance, recommendation, familiarity with the work or from previous experience with the building owner. In certain circumstances only one contractor may be able to provide the service required, eg. a repair to a specialist piece of equipment. Negotiation is concerned with the agreement of contractual arrangements and price. In negotiating, the client will compare the contractor's price with a base of listed prices and standards to give a check against the market level.

Daywork

The builder will carry out work and is paid for labour, materials and plant at cost, plus predetermined percentages for overheads and profit. The lack of financial incentive to encourage a steady and economical completion of the job is a strict disadvantage of the system. The maintenance manager will also not know entirely the cost of the operation until after completion. Therefore, it should be limited to urgent emergency works which cannot be priced competitively in advance.

The supervision of maintenance work

To ensure that the building owner obtains value for his money, maintenance work must be supervised. Inspections and checks must be undertaken to make certain that standards of workmanship and cleanliness are satisfactory, that materials comply with the specification and that jobs are completed fully. When work is being carried out in occupied premises or externally in or above areas where staff or members of the public have legitimate access, the maintenance manager has a duty in law to provide for the safe passageway of those persons and to ensure that they are not subject to hazards or dangers arising from the works in progress.

On large contracts the in-house or consultant surveyor responsible for the project may have an assistant who carries out a daily inspection of all work in progress.

Small contracts and works carried out on a quotation basis etc require interim inspections by the in-house manager of the building or one of his assistants to ensure that it is proceeding correctly and with a further inspection on completion. If work is being covered up by later operations, inspections must be carried out while it can be seen.

Where work is being valued on the basis of a schedule of rates as part of a term contract, periodic checks on measurements will be necessary as well as checking the workmanship and materials for compliance with the specification.

Service contracts require checks to be carried out whilst work is actually in progress otherwise doubts can arise as to whether operations charged have actually been performed, eg. the flushing or jetting of drains.

If work is carried out on a daywork basis, checks should be carried out whilst work is in progress on the number and trades of operatives employed. Plant utilised and any necessary standby plant for emergency use (eg. pumps for dealing with burst pipes or water seepage) should also be noted. Starting and finishing times of operatives should be monitored.

SELECTION OF CONTRACTORS

Lists of contractors should be compiled for the various categories of maintenance work who can be called upon to provide competitive tenders or quotations or execute emergency works on a call-out basis. Personal recommendations and formal references from other clients will assist in compiling lists. Categories suitable for maintenance work include:

(a) the small or medium sized contractor who undertakes at least 50% of his turnover in maintenance, improvement and refurbishment work. He will probably employ his own labour in the basic trades such as brickwork, carpentry and joinery and sublet the remainder to a number of regular sub-contractors;

(b) the small works department of a medium or large sized company. This type of organisation may have advantages for the client who has a number of depots or branches in different locations for it may be able to undertake a number of contracts simultaneously and be more willing to travel longer distances;

(c) a property maintenance company. The operatives, supervising staff and management of well run property maintenance companies have a specialised knowledge of the problems encountered with building maintenance. Usually this category of contractor will give a reliable service at reasonable cost;

(d) the jobbing builder, usually run by the sole proprietor carrying out most types of maintenance on a small scale as a general all rounder. Having very low overhead costs he may be able to provide a flexible service at low cost. Care must be taken to ensure that he is not under-capitalised and can provide the finance for all the resources necessary for his work;

(e) specialist contractors provide a very valuable service when electrical, heating, air conditioning or refrigeration plant etc, needs renewal, or where reroofing, redecoration or repairs to concrete need to be carried out. Often such services require small amounts of work from other trades. It is important to ascertain whether the specialist will cover these items, or whether the client will be required to provide attendant labour to the specialist. It may still prove more economical to employ a specialist rather than a general contractor in such cases as a general contractor may well employ a specialist for 95% of the work, requiring an additional layer of supervision and profit for two companies to be included in the tender sum.

(f) in addition to the specialist services indicated above there are national and local specialist contractors who provide emergency sub-trade services on a call-out basis for such tasks as the clearance of drain blockages, locksmithing, replacement of plate glass to shop fronts etc. Most of these will also undertake service contracts.

WORK BY DIRECT LABOUR

With efficient management the establishment of a direct labour force can be an economic way of carrying out maintenance work, provided that the workload is adequate for the continuous employment of operatives. With the exception of marketing the direct labour manager requires similar resources to a contractor. In addition to general supervision and administrative support, provision must be made for buying, estimating, work planning, measuring, costing, quality control, control of plant and financial management.

Procedures must be established to enable costs to be compared with those of contractors.

Quality of work should also be compared with that of contractors, although this may be difficult unless contractors are employed on similar works.

Since the introduction of the Local Government Planning and Land Act 1980 local authority building maintenance departments have been required to compete for the majority of their workload. The schedule of rates has become the most widely used basis of tendering for day-to-day repairs and maintenance in the public sector.

The legal requirement to compete and prove viable by achieving a stipulated saving on a trading account basis, has led to a general improvement in the efficiency and profitability of local authority direct labour organisations.

The need to achieve client satisfaction has also become a major consideration and this consideration is just as important with a company employing a direct labour force.

There are other circumstances where the employment of limited direct labour should be considered, even though the actual cost of the work by this method may be greater than employing contractors.

The following are examples of such circumstances:

(a) where instant action is required to a building or service to avoid loss of production or cessation of a vital activity of the business or organisation;

(b) where instant action is required to avoid the loss of valuable stock which must be kept at a constant temperature;

(c) where security is of paramount importance and positive vetting must be carried out before operatives are allowed to enter the area where maintenance is required.

CONCLUSIONS

It is essential that the building owner with his maintenance manager formulate a policy for the execution of maintenance work.

The policy should include:

● Which categories of work, are to be carried out by direct labour if a work force exists or if it has been decided to establish one.

● The financial limits for work executed by contractors on the following basis:

> daywork;
> quotation;
> term contracts;
> lump sum tenders.

● The response times acceptable for the various categories of maintenance work. If term contracts are entered into or direct labour maintenance works are executed on a schedule of rates basis, response times must be entered against each item on the schedule.

A separate schedule of items for emergency and high priority works carried out on a daywork basis can be given agreed response times.

Policy meetings should be held at pre-determined intervals when the desirability of changes can be discussed.

Chapter 8
Conservation

DEFINITIONS

The *conservation* of a building is that action which has to be taken to prevent undue deterioration and decay in the structure, the fixtures and fittings, which would affect the character and occupancy of that building.

Preservation, on the other hand. being largely aesthetic, concentrates on keeping a structure in its existing state–even if in a ruinous condition–and preventing further damage and deterioration.

Restoration, leans towards replacing damaged or missing parts of a building with new items designed to be copies of the original and intended to integrate harmoniously with the whole, with a pronounced imitative content.

Rehabilitation generally incorporates some degree of alteration or adaption in addition to repair and renewal elements and in its most extreme form may result in almost complete rebuilding.

The principles of conservation are largely applied to buildings which are of historic or aesthetic importance and include most ecclesiastical structures.

As conservation generally implies a greater degree of maintenance expenditure of a continuing nature than would be required for a building of less importance, a great deal of thought and planning has to be employed in ensuring the most effective use of the physical and financial resources available.

PRELIMINARIES

The agencies causing deterioration and subsequent decay in an older building are the weather, the environment, human factors, (vandalism or misuse), the force of gravity, chemical and biological attack from fungus or insects.

Before deciding on the courses of action needed to arrest deterioration one must first consider the original design, its structural advantages and weaknesses, undue wear or damage caused by human agencies and lastly the natural elements and their impact.

In the UK too scant regard is often paid to the effects of gale force winds and snow.

During the process of conservation work there must be no interference with any historical evidence which may be in the building or on the site. Therefore, before commencing such work it is necessary to carry out a full survey of the building to include materials and designs used in the construction as well as the present condition of the property.

From the survey report the 'degree of intervention' required to carry out the conservation work may be determined viz.

simple prevention of further deterioration;

preservation in existing condition;

structural consolidation;

restoration of structure, fittings and finishes;

rehabilitation, including possible improvement elements;

reproduction–imitiative or substitution;

reconstruction in whole or in part.

Generally, conservation work demands the services of highly skilled and experienced craftsmen who have a long lifetime of experience of traditional methods and materials. The choice of labour is particularly vital and consideration should be given to providing special training.

STRUCTURES, MATERIALS AND TECHNIQUES

Older buildings requiring conservation are generally less sophisticated and more massive than today's structures. Generally, weaker materials were used and 'rule of thumb' was the order of the day, the tendency being to ensure stability by over-designing with generous use of material resources. Another reason for the provision of strength by mass and sheer size was the intended permanence of structures.

The first thoughts of the conservationist must be directed towards the loadings on the different parts of the structure and, in particular, the foundations. Ground vibrations caused by road traffic can cause foundation displacement with resultant movement in the upper structure.

In addition to distortion of the structure caused by stress and load problems, settlement and distortion may be caused by the action of materials used. These can be broadly divided into defects which occur in the first phase of a building's life–say 25 years from completion–and those which take place at a later date or even have a cumulative effect over the years.

Deterioration or initial shrinkage of materials such as mortar may be responsible for initial deformation of the structure but longer term effects are confirmed by weathering, internal or external severe temperature change, the actions of salts (e.g. sulphate attack or efflorescence), rising damp and changing moisture content in the materials used.

Cracking in materials is a regularly occurring phenomenon in all old buildings. Superficial cracking occurs in applied finishes such as rendering and plaster but the more serious faults are caused by cracking in the structure itself, eg, cracks caused by settlement in the structure or subsidence which may be manifested as raking cracks in a brick wall, initially following the lines of the horizontals and perpends on the mortar joints but eventually cracking the bricks themselves.

The environment, in all its aspects, must be closely studied to assess the effect of all these factors on the building before any attempt at effective conservation can be made.

Local techniques of construction must be studied as it is most important that the reason why a building has been designed and built in a particular way out of certain materials should be appreciated before restoration or maintenance work is carried out.

When carrying out repairs, traditional materials compatible with the locality should be used wherever possible. Substitutes should only be used as a very last (and extreme) resort where no other means are available.

PLANNED MAINTENANCE

Effective and regular planned maintenance work is a major factor in conservation. It is the only logical way to ensure the retention of a historic or unique building so that it remains in a usable condition and does not deteriorate.

Planned preventive maintenance carried out to a regular schedule will prove far less expensive and inconvenient than allowing an older building to fall into such a state of disrepair that extensive and expensive emergency works become essential. This requirement for regular planned inspection and preventive maintenance is all the more important as materials and labour used in restoration work caused by neglect are normally far more expensive and difficult to obtain than those used in a more modern building.

Planned inspection to a routine can also have other uses by providing invaluable feed-back on performance over a very long period of time, often of materials and forms of construction which are no longer in regular use.

A planned maintenance system for a building or group of buildings also presents the opportunity in some cases to keep a small, directly employed labour force of skilled specialist tradesmen taken from the older and more experienced craftsmen.

Neglect of an old building in partial or constant use can be mitigated by briefing staff, caretakers and in particular cleaners about the tell-tale signs of wants of maintenance.

Another highly valid reason for using all eyes, hands (sometimes noses) available from the regular staff in a historic or valued building is that such properties are not, as most modern buildings, constructed to last for a limited life but are expected to be of a permanent nature and, therefore, any fungoid infection or insect infestation problems must be reported promptly before cumulative damage can occur. However, when taking remedial measures it is best to keep to the same methods and materials as were used in the original construction, wherever possible, as the introduction of new 'foreign' materials can sometimes have a serious and even catastrophic effect.

Plumbing, electrical and heating services all have a limited life and this is a factor which demands particular attention in a planned maintenance scheme. Wires, pipes and radiators are best painted in to match the surfaces on which they are mounted and replacement light fittings should be carefully chosen and sensitively sited.

Although some abandoned buildings and ruins are merely to be kept in either a wind and watertight or structurally safe condition, most buildings to which a conservation policy applies require more sophisticated treatment in the form of a quinquennial plan covering major repairs, pointing and decoration with regular clearing out of gutters, gullies, drains, valleys and parapet gutters in particular–say twice yearly. Regular routine inspections of security and fire alarm systems, smoke and heat detectors and emergency lighting, together with appropriate testing, are very necessary in the larger or more important buildings.

This 'acceptable standard' for conservation purposes is often dependent upon the amount of finance available. Lack of maintenance funding in many cases means that very careful pre-planning of resources available must be carried out to include a strong element of preventive maintenance to guard against expensive, catastrophic failures, as well as a strict system of priorities. This should ensure that, for example, structural and roof repairs as well as external painting and gutter maintenance have priority over internal decorations and work of a purely cosmetic nature. If such a list of priorities is compiled and worked through as the funds become available, then this should ensure that essential repairs are not neglected.

The cost and scope of the planned maintenance works–be they preventive, corrective or emergency–must be recorded accurately and in detail so that not only can cyclical works be checked and not forgotten but the performance of the structure, its fixtures and fittings can be monitored continuously. In the case of a 'permanent' building, such records become an essential guide to those dealing with the maintenance in future years.

Access for maintenance purposes is often a problem in older buildings. Walkways and access hatches in roof voids, permanent access and walkways to high-level roofs, hatches to inspect and repair sub-floor areas as well as ducts and accessible trunking for services will all help to reduce future costs and ensure that no part of the structure remains neglected. The more inaccessible the area of a building to be maintained the more essential it is that the repairs should be carried out with a longer-term cycle in mind, even if the cost is greater at the time.

Other than the stress to be placed on the special features for old, large historic buildings, the pattern of a planned maintenance programme should fall broadly into the same organisational process as for any other building. This will be based on a quinquennial cycle for major repairs, restoration and detailed inspection and report with preventive work and major repairs as well as cleaning based on a daily, weekly, monthly, quarterly and annual maintenance to cover rainwater and soil drainage systems, roofs, gutters, downpipes, doors and windows, fittings and services.

Appendix A

TWO ORGANISATIONS USING COMPUTERS IN BUILDING MAINTENANCE

NATIONAL HEALTH SERVICE (NHS)

The Department of Health and Social Security (DHSS) has been using computers in building maintenance in the NHS for many years and began the development of its WIMS (Works Information and Management System) suite of packages in 1979. It has seen a tremendous development in computing technology as well as having had the benefit of many 'user-years' experience in the application of the software. The original software is now referred to as WIMS 1 because a new set of applications is now being designed under the title of WIMS 2.

The WIMS 1 suite comprises 13 different modules, each designed to assist in a specific area of estate management in the NHS. The modules are:

- Asset management
- Stock control
- Energy monitoring
- Redecoration
- Budget monitoring
- Maintenance contracts (linking asset management and budget monitoring)
- Property appraisal
- Property management
- Annual maintenance plan
- Residential property
- Contract management (capital projects)
- Electro-medical equipment management
- Vehicle management

Over the years, the most widely used modules have been asset management (used mainly at individual site/hospital level), property management (at District and Region) and budget monitoring (used at all levels). The property appraisal module is finding increasing application because of the current rationalisation of the NHS estate.

WIMS 2 is being developed in a more integrated form, although it is not feasible to produce a single massive integrated system covering all NHS estate functions. It is being designed around four applications:

- Asset Operational–providing an asset register with forecasts of replacement costs and historical cost records;
- Property and Land–includes a property register, property appraisal facility and an ability to access costs on individual properties from the Asset Operational application;

59

- Estate Strategic Plan–covers long-term costing, scheduling and scheme progress information for work involving capital and revenue expenditure;
- Financial Control–deals with budget monitoring, stock control and purchasing, invoicing and job costing.

The WIMS 2 applications make use of the latest report generation facilities available with the PICK and UNIX operating systems under which they are being implemented. The implementation philosophy ensures that the system will be fully documented and capable of being mounted on any combination of hardware and software with relative ease.

THE LONDON BOROUGH OF WALTHAM FOREST

After running a 'first generation' repairs reporting system for a number of years, the London Borough of Waltham Forest commissioned a software firm to write a suite of programs to deal with all its housing repairs and maintenance systems. The results is an on-line and real time system providing access for over 150 housing staff to the following applications:

Property Database

This includes records of individual dwellings, garages and blocks. Each property type contains details of more than 40 separate elements. Individual records may be updated via the computer keyboard, but mass input can be derived from condition survey data using optical mark reading devices.

Works Orders

Orders for repairs are raised directly on terminals by repairs reception clerks using a schedule of rates to specify the work. The system includes a repairs history and holds planned maintenance indicators against properties where major repair work is either under contract or imminent. A series of reports can be produced to show jobs completed, overdue, or to record any other measure of performance.

Contracts

This module enables contracts for general building work, servicing, repairs to relets, external painting etc, to be recorded and their progress monitored against time and cost. Links to the property database indicate which properties are included in particular schemes.

Financial Control

The system contains full provision for invoicing, approvals for payment and budgetary control. Expenditure may be monitored by scheme, property type, trade, or any combination of similar variables. This permits a very flexible approach to be adopted in the use of cost reporting centres.

One of the features of the system is its ability to communicate with other systems within the authority, for example with the DLO computer, and the mainframe, which runs the Housing Department's Rents and Housing Benefits systems. This enables screens connected to the repairs and maintenance system at remote local estate offices to be used for interrogating and updating rent records.

Appendix B

ESTIMATED LIFE EXPECTANCY-THE LIFE EXPECTANCY OF BUILDING MATERIALS

THE POTENTIAL USES OF LIFE EXPECTANCY THEORY

A knowledge of the life expectancy of building materials and components is important both in the design and the maintenance of buildings.

Initially, at the construction stage, the client's brief should clearly state the required life of the complete building and also its constituent parts. This will enable the designer to select the products that best satisfy these requirements within the accompanying constraints of cost and quality. These latter considerations will help to decide whether the required life is best achieved by using durable components-which last as long as the building itself-or by using less durable, normally cheaper, products that need replacing more than once during the life of the building. As a simple example, an expensive wood block floor, properly maintained, should give a life of 60 years, compared with much cheaper vinyl tiles that will need replacing at least twice within the same period.

It is up to the designer to make such decisions and also to acquaint the client with the future maintenance and refurbishment consequences of those decisions, preferably by preparing a fully costed model of future repair and replacement needs. The model would also be required if the client wished to use the building audit method of identifying future improvements.

However, few new building owners are in the fortunate position of having such a clear indication of their future maintenance commitments, and it is left to their maintenance or facilities managers to fill the gap, by constructing their own versions of such models. In such instances, particularly where as-built drawings and maintenance records are poor, it will usually be necessary first to identify what materials have been used in the construction of the buildings and then their current condition, by means of fairly extensive condition surveys. Once these data are available, it should be possible to draw up a costed programme of repair and replacement work using life expectancy estimates in a similar way to the construction of the original design model referred to above.

AVAILABILITY AND USE OF LIFE EXPECTANCY DATA

Regrettably, both for the designer and the maintenance manager, very little guidance exists on estimating the life expectancies of building products, even for some of the most familiar items like kitchen units and sanitary ware. The former Scottish Special Housing Association (SSHA) prepared its own schedule of average lives for assessing the cost of future programmed repairs and replacements, and this principle was adopted by the Audit Commission in its predictions of future replacement costs for the public sector housing stock as a whole in its report *Improving Council House Maintenance* published in 1986. As part of the same study, a booklet was published in 1985 which compiled all the sources of data on life expectancies and maintenance cycles then available. Besides bringing together a wealth of relevant material into one publication this underlined the importance of regular maintenance in prolonging the useful life of building elements and components, by a combination of periodic inspections and condition assessments.

However, it has been left, to a recent draft British Standard* to clarify the important distinction between permanent and renewable parts of a building, while providing for both to be included in the long term maintenance plan. Some parts of a building are clearly intended to last for its whole life, such as the main structure and floors, although they may or may not require some partial *maintenancess* during that time, whereas other components that are intended or known to requires *renewals* at least once during the building's required life. This has been an obvious shortcoming of models that make the assumption that every element has a finite life. They ignore, for example, that most types of brick wall will last indefinitely if repointed at the appropriate intervals.

The technique proposed by the British Standard can be used to show both maintenance and renewal needs for a range of components that might be used in house construction. It identifies building material within component, element and location in a format based on that used by SSHA; shows whether items are intended to be permanent (P), maintained (M), maintained prior to renewal (M/R), or simply renewed (R); and gives a fairly straightforward process to construct a model of the future maintenance costs either of a new development, or of an existing estate, given the necessary assessment of its existing condition.

The life expectancies indicated are not definitive, but are averages drawn from a number of reliable sources, including those already mentioned. This is an area where much more work needs to be done, but maintenance managers can do much to help themselves by keeping records of what is happening on their properties and by exchanging information with their peers in organisations with similar property portfolios.

---

* A revision to Chapter 9 of CP3: 1950 Durability which is scheduled for publication in 1990.

Appendix C

MAINTENANCE SERVICE STANDARDS

	Operation	Frequency	Relationship to other operations or time allowed to complete operation
1	Repairs prior to painting *(external painting)*	5 years	Inspection 1½ years in advance of painting. All major work to be included in estimates, and executed 1 year ahead of external painting; minor repairs three months before painting. All external hinges, tee bands to be lubricated every 5 years.
2	*Painting and decorating (external)*		
	Normal external painting of all property	5 years	Including common staircases to flatted blocks with lifts.
	New property external painting	2 years	
	Short-life property external painting	5 years	To be programmed if life is two years or more and
	Stained woodwork	7½ years	it has not been painted for five years
3	*Painting and decorating (internal)*		
	Main entrances to all blocks of flats, including hallways and internal corridors together with staircases to blocks of flats without lifts	2½ years (or earlier on inspection if in poor condition)	To be included with external painting programme as wet-weather work and additional treatment to fit programme. Surfaces not painted to be cleaned down. All main entrances to flatted blocks to be inspected annually and treated immediately if in poor condition.
	Normal programme work (limited liability), if finance available, to the following:		All existing metal lift-off hinges (particularly rising butts) to be re-lubricated with grease at least every 5 years or during cyclical redecoration of the premises. Cylinder locks/latches to be re-lubricated with light oil.
	(a) house or maisonette	5 years	Hall, landing or staircase only
	(b) flat or bungalow	5 years	Hall, and any landing and staircase together with one bedroom.
	(c) sashes and frames	5 years	Throughout the building if necessary to preserve the fabric and prevent wet rot or decay.
	(d) special cases	5 years	Restricted to OAPs, one-parent families or disabled persons, and covering minimum redecoration to all occupied rooms.
	Empties – estate properties at change of tenancy		Repair all work necessary to rooms being redecorated. Check window safety catches, oil or grease all hinges. Check all rising butts and oil and treat as agreed. Renew where burring has occurred on splay-bearing surfaces. Work to be completed within 3 weeks of receipt of keys.
	Empties – general properties	3 weeks	Repair all work necessary to rooms being redecorated. Check window catches, safety catches, oil or grease all hinges. Treat rising butts as above.
	Common staircases to flatted dwellings without lifts	2½ years (approximately)	To be included with external painting programmes as wet-weather work. Non-painted surfaces to be washed down.
	Laundries	2 years	On average, including machinery.
4	*Day-to-day repairs and servicing*		
	Emergencies (safety, security, health reasons)		Complete within 24 hours of notification to management, or make safe if a long job. (Progress must be checked daily by management until completed.)
	Bricklaying repairs		5 weeks average programme, but no repairs to be outstanding more than eight weeks.
	Carpentry repairs		4 weeks average programme, but no repairs to be outstanding more than six weeks.
	Plumbing repairs		3 weeks average programme, but no repairs to be outstanding more than five weeks.

Glazing renewals	2 weeks average programme, but no repairs to be outstanding more than four weeks.
Electrical repairs	Average programme, but no repairs to be outstanding more than three weeks.
Space heating	Average programme.
Miscellaneous repairs	5 weeks programme.
Fencing repairs	5 weeks programme.

	Operation	Frequency	Relationship to other operations or time allowed to complete operations
5	*Servicing of appliances and equipment etc.*		
	Gas appliances including boilers	Annually	Only specially trained staff to be engaged on this work. Central heating units to be serviced during summer period, hot water heaters all year round. Special attention to be given to ventilation.
	Electric heating (all types)	Annually	To be treated as emergency repairs and dealt with immediately
	Immersion heaters	Annually	To be treated as emergency repairs and dealt with immediately.
	Vent Axia fans	Annually	To be cleaned, serviced and tested.
	Door closers	Annually	To cover complete range of door closing equipment.
	Soil stacks	Annually	To be thoroughly cleaned. Back surges to be reported to management for attention to drains.
	Scale reducers	Annually	Or more frequent where required.
	Cleaning of ventilation ducts and individual extractor fans	Annually	During summer period.
	Gulley and drainage channel cleaning	Annually	To include all public and private balconies.
	Petrol and oil interceptors	Annually	To be inspected and cleaned out where necessary.
6	*Inspection and tests*		
	Roof storage tanks	Annually	Special attention to be given to covers, overflows and traps.
	(a) Multistorey blocks externally and common parts internally	5 years	To be thoroughly inspected 18 months ahead of external painting and followed by preparation of programme of works to meet requirements of item 1. Any Defective Premises Act or Health and Safety at Work Act matters to be reported and dealt with immediately. Blocks over 6 storeys in height to be inspected from cradles.
	(b) All other property externally		
	Power tool inspection and tests (including portable hand tools)	(a) Daily inspection when in use	Immediate attention to be given to any defects. Defective tools shall not be used.
		(b) Monthly and annually	To be inspected by the manufacturers or their agents at intervals not exceeding six months, and any defect corrected immediately. Unserviceable machines not to be used under any circumstances.
	Scaffold and plant	Monthly	Plant for external painting serviced in winter period and for internal painting to be serviced in summer period. All to be inspected monthly and appropriate action taken to ensure that only scaffold and plant in good condition is available for use.
7	*Refuse*		
	Cleaning of hoppers on staircases and access balconies	3 times each year	Check to ensure hopper seal working correctly, check ventilation and that the chute lining is in a sanitary condition. Any grease or other dirt to be removed by approved treatment.
	Clean dust chute chambers	3 times each year	Paint internally or wash down as specified. Any defects to be reported to management for action.
	Clean chute cut offs	3 times each year	Repair immediately any defects or renew cut off.
8	*Horticultural work*		
	Grass cutting	Every 9 working days in normal growing periods	Reduce to 7 working days or less in peak growing periods, or increase to 10 working days in slow growth periods of drought.
	Hedge cutting	Twice a year	April/May and August/September. Cut only once in the year following splitting.

Hedge splitting to 750 mm high × 250 mm wide	1/5 each year	5 year programme. Between October and March.
Winter programme	Yearly	

9 *Inspections and servicing of mechanical plant, equipment and laundry equipment to comply with Health and Safety at Work Act 1974 and other appropriate legislation*

Laundries and drying rooms		
(a) Check mechanism of washing machines, hydro extractors, spindryers, and tumble dryers	Quarterly	Resulting work to be programmed for immediate attention. Defective equipment must be made inoperable by withdrawing the fuses, locking the rooms, turning off gas or other services, disconnecting or capping off if necessary.

Operation	*Frequency*	*Relationship to other operations or time allowed to complete operations*
(b) Examine and test electrical circuits of washing machines, hydro extractors, spin dryers, and tumble dryers	Quarterly	Any resulting work to receive immediate attention.
(c) Examine and test electrical circuits of electrically operated drying cabinets	Monthly	Any resulting work to receive immediate attention.
(d) Examine and test operation of gas-fired drying cabinets, and ensure adequate ventilation and satisfactory working of safety devices	Monthly	Any resulting work to receive immediate attention.
Woodworking machinery (including timber treatment plant, compressed air plant and equipment, dust extractor plant) and all other machinery	(a) Daily inspection (b) Not exceeding 6 months period	Give immediate attention to any defect or servicing required. To be inspected by the manufacturer or his agents at intervals not exceeding six months. Correct any defect immediately. Unservicable machines not to be used under any circumstances.

10 *Playground equipment*

All playground equipment	Monthly	All playground equipment to be tested monthly and action taken immediately on any defects.
Exhaustive condition tests on playground equipment	5 yearly	Equipment found to be defective must be immediately immobilised and made safe, followed by immediate action to repair.

11 *Dry risers and all fire-fighting*

	Twice per year	Check dry risers to ensure that all valves and fittings are in position and undamaged. Check all hose reels and fittings and all fire extinguishers to ensure that they are fully charged and securely fixed. Resulting work to be programmed immediately. Arrange for fire prevention officer to check at the appropriate intervals.

12 *Electrical re-wiring (renewals)*   20-30 years

13 *Renewal of gas water heaters*   25 years

CATEGORIES OF BUILDING MAINTENANCE

Various definitions of maintenance have been drawn up over the years, some designed to apply specifically to building maintenance, others, such as BS 3811:1984, using terminology intended to embrace both building and engineering plant maintenance within the wider field of terotechnology.

All language is subject to development and change, but building maintenance in particular–as it emerges into prominence as a subject in its own right–has tended to develop a terminology of its own. The following definitions, while acknowledging previous contributions on the subject, are intended to reflect existing custom and practice throughout the industry.

Building maintenance

For building maintenance in general, The Chartered Institute of Building supports the following definition:

'Building maintenance is work undertaken to keep, restore or improve every facility, ie every part of a building, its services and surrounds to an agreed standard, determined by the balance between need and available resources'.

Within this overall definition, building maintenance can be sub-divided into a number of different types of work, each of which may be used by a building owner to fulfil a particular purpose during the life of a building and for which specialised systems and practices may be identified. The four principal types are:

● jobbing, or responsive repairs;

● cyclical maintenance;

● planned maintenance, or programmed repairs;

● refurbishment or improvement works.

Jobbing, or responsive repairs

Jobbing repairs are repairs of a minor nature that are carried out 'on demand' in response to the requests of tenants or occupants of properties, usually with people in occupation of the premises and within a fairly short time scale of the request (eg between 1 and 28 days of notification). The work ranges from emergency call-outs caused by serious electrical or plumbing problems to straightforward jobs like replacing defective gutters. They are also referred to collectively as 'day-to-day', 'routine' or 'reactive' maintenance.

Cyclical maintenance

Many people are acquainted with cyclical maintenance in the form of regular annual visits by the area gas boards or contractors to service the central heating systems in their homes. It is essentially preventative maintenance, undertaken at regular intervals to parts of a building or its services in order to minimise breakdowns and preserve the integrity of the complete unit.

Therefore, it is usually applied as a matter of policy to items such as central heating, fire alarms, lifts, security installations and also external paintwork, where regular re-decoration plays a vital part in preserving the life expectancy of timber and metal components.

Planned maintenance or programmed repairs

Planned maintenance is increasingly used to describe the more extensive repairs or renewals that are undertaken as part of a programme of works to restore major elements or components of a building to acceptable condition and working order. The programme will have been drawn up as the result of regular condition surveys designed to establish acutal need by balancing the remaining life of the building fabric, services and finishes, against economic and social criteria. This is a somewhat different approach from the engineering context, where planned maintenance is more likely to mean the replacement of parts of items of equipment before they reach the end of their useful lives, where there could be a risk of serious mechanical failure or breakdown.

Refurbishment or improvement works

As the overall definition implies, maintenance is not simply a mixture of 'running repairs' and replacing like with like when individual components wear out. Periodically, it will be necessary to invest in refurbishment or improvements in order to maintain a property to a standard appropriate for its intended use or to bring it into line with current standards and thereby maximise its asset value. This is particularly true of the commercial sector, but applies also to rented housing, where tenants' expectations increasingly reflect the standards that apply to the owner-occupier, for example, in respect of items such as replacement windows and sanitary ware.